WATER PRODUCTIVITY AND FOOD SECURITY

GLOBAL TRENDS AND REGIONAL PATTERNS

Current Directions in Water Scarcity Research

VOLUME 3

Series Editors

Robert Mcleman

Michael Brüntrup

WATER PRODUCTIVITY AND FOOD SECURITY

GLOBAL TRENDS AND REGIONAL PATTERNS

M. DINESH KUMAR

Institute for Resource Analysis and Policy,
Hyderabad, Telangana, India

ELSEVIER

Elsevier
Radarweg 29, PO Box 211, 1000 AE Amsterdam, Netherlands
The Boulevard, Langford Lane, Kidlington, Oxford OX5 1GB, United Kingdom
50 Hampshire Street, 5th Floor, Cambridge, MA 02139, United States

ISBN: 978-0-323-91277-8
ISSN: 2542-7946

For information on all Elsevier publications
visit our website at https://www.elsevier.com/books-and-journals

Publisher: Candice Janco
Acquisitions Editor: Louisa Munro
Developmental Editor: Michelle Fisher
Production Project Manager: Kumar Anbazhagan
Cover Designer: Mark Rogers

Typeset by Straive, India

Working together
to grow libraries in
developing countries

www.elsevier.com • www.bookaid.org

Dedication

This book is dedicated to my late father, P. O. Kumaran Nambiar.

Contents

Preface

The idea of this book came sometime during 2019 when I started reading some research articles providing estimates of water productivity of irrigated crops in some of the most water-rich regions of the world. These articles argued the need for improving water productivity in highly water-intensive crops like paddy to benefit the farmers and also save water for society. It appeared as though these researchers considered water productivity improvement measures as "benign" and that such measures could help both farmers and the society at large, irrespective of where those measures are introduced. These studies ignored the importance of considering the relative scarcity of land and water in the region or locality under investigation prior to setting the research goals. The cardinal principle is that water productivity improvement measures should be a concern for both agriculturists and the society at large in areas where cultivable land is abundant and water is scarce.

In fact, during the last 10 years or so, the ideas such as "more crop per drop" of water, "water-saving in agriculture and reallocation of water to other sectors of the economy through the water use efficiency route," etc. dominated the water management discourse in several developing countries of South Asia and Sub Saharan Africa. The water policymakers in these countries are receptive to those who espouse these ideas, without making any efforts to understand the nuances in terms of what it means for the farmers, the governments, and the society. The public expenditure in countries like India, Pakistan, Afghanistan, and Sri Lanka on water-saving systems (such as drips and sprinklers) and decentralized energy production systems based on solar photovoltaic (PV) system to run these systems is going up. The enthusiasm among irrigation bureaucracies of these countries to invest in the lining irrigation canals to save the "water lost in conveyance" is also unabated.

As research from several arid regions of the world suggests, what is fundamental is the fact that adoption of water use efficiency/water productivity improvements in crop production in many situations (where water is relatively scarce, and land is relatively abundant) can result in farmers diverting the saved water to expand the irrigated area, thereby saving no water, even if real water saving is achieved at the plot level. However, this is ignored by researchers, water professionals, and policymakers. What is also ignored is the fact that in many situations (where neither water nor electricity for pumping water is priced), farmers do not gain economically by improving the productivity of water use in irrigation, and on the contrary, could lose due to yield reduction. The outcome of such measures for agricultural outputs and food security can be negative if farmers do not have land to expand the area under cultivation. Overall, the roles of institutions and policies in aligning the societal goal of water conservation and food security, and private interests of the farmers in maximizing farm income are least appreciated.

Obviously, there was a need to change the style of the discourse, and bring important elements such as economic viability, overall agricultural production, and food security and water conservation, land scarcity, etc. in the purview of scientific inquiry on agricultural water productivity improvements. I, therefore, decided to revisit some of the earlier works done on agricultural water productivity, update and build on them and also synthesize the works of others so as to discern some crucial messages for the global audience.

The work encompassed several disciplines such as plant physiology, agronomy, water use hydrology, river basin hydrology, agricultural economics, food trade, and institutions and policy. More painstaking was the task of collecting and analyzing data for other regions of the world, especially Sub-Saharan Africa and East Asia. When I first approached my esteemed colleague from *Elsevier*, Ms. Louisa Munroe, who is their Commissioning Editor, with a short outline of the proposed book, she expressed great interest in it and asked me to submit the full proposal. This was promptly done. The responses received from the reviewers were very positive, constructive, and highly encouraging. Working with Louisa and Editorial Project Manager Madeline Jones was a very pleasant experience throughout.

I sincerely hope this book "*Water Productivity and Food Security: Global Trends and Regional Patterns*" would elevate the level of the international debate on several issues such as the technologies and practices, institutions, and policies required for achieving the multiple goals of water productivity, water conservation, food security, and farmer income; and the need to improve the rigor in the research on water productivity and food security in agriculture the water sector.

M. Dinesh Kumar
January 25, 2021

Acknowledgments

First of all, I am extremely thankful to the four reviewers, who had gone through the book proposal, and the editorial office of Elsevier Science, which immensely helped me in framing the "problem" which the volume addresses, sharpening the focus of the contents especially that on Sub-Saharan Africa and enhancing the value of final manuscript. All the reviewers were quite meticulous in their observations and I am very glad to accept most of them.

I extend my gratitude to my colleague, Dr Saurabh Kumar, Researcher, Institute for Resource Analysis and Policy, for doing an excellent literature review digging out the latest scientific articles and reports on crop water productivity and climate change and synthesizing them. I used them in the fourth chapter of the book. I am grateful to Mr Ajath Sanjeev, my Executive Assistant who meticulously checked and formatted the references and prepared the table of contents, and list of tables, figures, and abbreviations of the book.

I am also extremely grateful to my beloved daughter, Archana Dineshkumar Manhachery for meticulously editing the entire manuscript for language and style.

I owe profusely to my beloved family, especially my wife Shyma and my daughter who stood by me during the difficult times, provided the moral strength and became a constant source of motivation.

Lastly, I proudly dedicate this book to my beloved father, late Shri P.O. Kumaran Nambiar, who left for the heavenly abode 44 years ago at the age of 40, and whose fond memories as a loving father and teacher still drive me to work with fire and passion.

M. Dinesh Kumar
January 25, 2021

Abbreviations

AWD	alternate wetting and drying
AWR	annual water resources
BCM	billion cubic meter
CA	cultivated area
CAGR	compound annual growth rate
CCCM	Canadian Climate Central Model
CERES	crop environment resource synthesis
CF	consumed fraction
CGWB	Central Ground Water Board
cm	centimeter
CO₂	carbon dioxide
CU	consumptive use
CWC	Central Water Commission
CWP	crop water productivity
DSSAT	Decision Support System for Agrotechnology Transfer
EGP	Eastern Gangetic Plains
ET	evapotranspiration
FAO	Food and Agriculture Organization of the United Nations
GAEZ	Global Agroecological Zones
GBM	Ganga, Brahmaputra and Meghna
GCM	General Circulation Model
GDP	gross domestic product
GIS	geographic information system
GISS	Goddard Institute of Space Studies
GLAM	general large-area model
GLYCIM	soybean simulation model
GOO	Government of Odisha
GOP	Government of Punjab
ha	hectare
HadCM3	Hadley Centre Coupled Model version 3
HDI	Human Development Index
IBIS	Indus Basin Irrigation System
IBSNAT	International Benchmark Sites Network for Agrotechnology Transfer
ICASA	International Consortium for Agricultural Systems Applications
ICID	International Commission on Irrigation and Drainage
IGNP	Indira Gandhi Nahar Project
IPCC	Intergovernmental Panel on Climate Change
IPM	Integrated Pest Management
IWMI	International Water Management Institute
kg	kilograms
LAI	leaf area index
m	meter
MCM	million cubic meter

MI	microirrigation
mm	millimeter
MSM	Maize Stover Mulch
NRDP	nonrecoverable deep percolation
NRM	Natural Resource Management
OECD	The Organisation for Economic Co-operation and Development
PE	potential evaporation
PET	potential evapotranspiration
PFP	partial factor productivity
PM	plastic mulching
PRECIS	Providing REgional Climates for Impacts Studies
RW	ragweed
SEBAL	surface energy balance
SSA	sub-Saharan Africa
SWAP	soil–water–atmosphere–plant
SWAT	soil and water assessment tool
TFP	total factor productivity
UNDESA	United Nations Department of Economic and Social Affairs
UNDP	United Nations Development Programme
UNESCO	United Nations Educational, Scientific and Cultural Organization
WAI	water adequacy index
WHP	white hoary pea
WP	water productivity
WUE	water use efficiency

List of Figures

List of Tables

CHAPTER 1

Water productivity improvements for agricultural growth and food security: Where and why?

1.1 Introduction

Widespread perception about growing water scarcity and an impending water crisis had prompted water resource scholars, water resource professionals, and policymakers to focus on improving the productivity of water use in agriculture, based on the premise that agriculture is the largest user of water globally, and improving water productivity in agriculture would allow nations and regions to save water that can be allocated to other sectors such as municipal and industrial use and ecology.

It is quite an established fact that the world is not homogeneous when it comes to water situation and that there are water-rich and water-abundant regions/countries along with water-stressed, water-scarce, and absolutely water-scarce countries (Falkenmark, 1989, 1997), and water productivity improvements would be essential in countries that belong to the latter three categories. However, this important consideration for choosing priority areas for agricultural research hasn't stopped researchers and practitioners from looking at water productivity enhancement as a universal option for enhancing agricultural outputs.

One reason for this misplaced priority is that historically, analysis of water scarcity problems and food security challenges confronting nations around the world treated them as a single challenge, using simplistic considerations of renewable water availability and extent of withdrawal of renewable (blue) water from the natural system (see Falkenmark, 1997; Kummu et al., 2016; Raskin, Smith, & Salt, 1997). Consequently, irrigation water availability is

Current Directions in Water Scarcity Research, Volume 3
https://doi.org/10.1016/B978-0-323-91277-8.00008-3

often used to indicate food security challenges, ignoring the contribution of land (Beddington, 2010; Falkenmark, 1997; Mancosu, Snyder, Kyriakakis, & Spano, 2015).

As a result, works on water productivity improvement has included regions that are food insecure, although most of those regions are not water-scarce, and as reported by Kumar, Bassi, Singh, Ganguli, and Chattopadhyay (2017), they are unable to produce sufficient food either because they have not developed all the utilizable water resources for expanding irrigated area or are acutely land-scarce.

Ideally, discussions on water productivity improvements should focus on regions where either water is becoming a limiting factor (either physically or in financial terms) for increasing agricultural production, or water resources are getting mined due to intensive use for agriculture with less amount of water available for economic activities and ecological functions. The discussions can leave out regions where water is abundant for crop production and land is a limiting factor. The discussion should also leave out regions where largely soil water is used for crop production and the agricultural surplus is because of a large extent of arable land being put to cultivation and animal grazing.

At the global scale, the focus should be on distinguishing regions that are really "water-scarce" from those that are "food-insecure." The focus of research and investments on water productivity improvements should be on the first category of region. This will help to save the precious resources from being spent on technologies (both crop and irrigation) for improving crop water productivity. Apart from the issue of finance, theoretically, many interventions to improve water productivity including deficit irrigation and adoption of certain shorter duration varieties come at the expense of yield reduction, which can create more problems for the food insecure regions that are land-scarce. Yield enhancement should be the immediate research priority in the second category, which includes countries like Bangladesh.

Within the second category of countries, i.e., "food insecure" countries, we need to separate those which are "water-rich" and "land-scarce" from those where available water resources are yet not fully developed to intensify cultivation in the available cropland, like some of the countries in Sub-Saharan Africa. The reason is that the latter might have to focus on water productivity improvements in crop production, along with investments in irrigation in the long run, as water resources get increasingly appropriated and water availability for ecological functions increasingly become scarce.

Another important concern in the area of agricultural water productivity, which is from the perspective of water resources management, is concerning the scale at which water productivity needs to be measured.

This concern stems from the fact that many of the measures that enhance crop water productivity at the field level may not result in water productivity improvements at the basin level, as they ignore the beneficial use of the water which gets recaptured and reused in one part of the basin as a consequence of deep percolation and/or runoff losses that takes place elsewhere (Ahmad, Masih, & Turral, 2004; Van Dam et al., 2006). In other words, water productivity assessment is "scale" dependent. It can be measured at various scales—from plant to plot to field to scheme to finally the river basin. Depending on the unit of analysis, the denominator of water productivity function could change (Molden, Murray-Rust, Sakthivadivel, & Makin, 2003).

While at the plant level, "transpiration" is used in the denominator for estimating water productivity (kg/T), at the plot or field level, the total water applied is what matters (kg/vol. of water applied). However, at the basin level, the total water depleted is what matters.

Because of these scale effects, the distinction is made between national water-saving and real water saving, with the latter, argued to be resulting from either reduction in consumptive use of water in crop production or prevention of loss of water into the natural sinks or both (Seckler, 1996). However, the controversy surrounding what technological measures can actually result in real gains in terms of water productivity improvements and real water saving at the basin level doesn't seem to be dying out (see Frederiksen & Allen, 2011; Frederiksen, Allen, Burt, & Perry, 2012; Gleick, Christian-Smith, & Cooley, 2011; Perry, 2007).

There were intense academic debates in the recent past on water productivity improvements in agriculture. They addressed the question of maximizing the benefits and minimizing the nonbeneficial component of the water depleted in a basin, but merely at a theoretical level. They did not explain under what conditions, in a "closed" basin, applied water-saving results in real water saving. As a result, these discussions have limited relevance for framing policies and designing programs to fit specific regional and local contexts. Also, the discussions are largely on crops, and not "farms" to which farmers allocate their water (Kumar & van Dam, 2013).

The third concern pertains to the over-emphasis of enhancing crop per drop of water in global academic research on agricultural water productivity. This can be explained by the fact that most of the research has been done by agricultural scientists who are concerned with increasing biomass production per unit of water consumed by crop in the form of evapotranspiration. But as pointed out by Kumar and van Dam (2013), while in many developed countries where water economy is formal (with the price for irrigation water and energy used for lifting and using water in agriculture priced based on consumption, and limits on volumetric water withdrawal), enhancing yield per unit of water made sense, the farmers in most developing countries are not concerned with raising crop yield per unit volume of water. Instead, they are concerned with maximizing return per unit of land and to an extent the net income per unit of water. The point to be reckoned with is that with no marginal cost of using water and energy, in the developing country context, for any crop, the irrigation corresponding to the highest yield and highest gross and net return can be far higher than that corresponding to the highest water productivity (Kumar & van Dam, 2009, 2013).

The real opportunities for enhancing WP and saving water from agriculture in a basin in the case of developing economies of Asia and Africa in arid and semi-arid tropics, need to be analyzed in terms of economic and social returns per unit volume of water depleted, as agriculture is the major source of livelihood for the smallholders; water is scarce at the societal level; and food security and rural employment are still major concerns there (Kumar & van Dam, 2009, 2013).

1.2 Purpose of this book

There are three distinct concerns about agricultural water productivity improvement in developing countries. They are (i) the regions where water is a limiting factor for raising agricultural outputs and water productivity improvements are necessary for enhancing agricultural outputs to prevent additional environmental water stress; (ii) the technological measures in irrigation that can raise agricultural water productivity and result in water-saving at various scales—from plant to plot to the farm to scheme to river basin—, and under what conditions they help conserve water resources, and (iii) the opportunities that exist in the developing economies of South Asia and Africa for raising water productivity and improving water economy at the basin scale, without compromising on the resilience of the farming systems, food security, and employment generation in the farm sector. These questions are addressed in this book.

The first part of this book analyses the food security and water management challenges of individual nations but does so by delinking the food security challenges from that of supplying water to meet the needs of industries, livestock, domestic and environmental sectors. For this, three indices were developed, viz., water adequacy index; water-land index, and a composite water-land-pasture land index, and their values were computed.

The second part of this book deals with water productivity in agriculture globally in general, and South Asian in particular. The first section, it deals with the conceptual differences between water use efficiency and water productivity. It then provides an overview of the international debate on the conceptual issues surrounding the definition of water productivity. It goes on to discuss the various dimensions of water productivity. It also takes a relook at the relevance of estimating water productivity in the light of total factor productivity.

The second section, to clarify issues involved in estimating water productivity, discusses the theoretical aspects of estimating production function in crop production, makes the distinction between "marginal productivity" and "average productivity" of water, and "technical efficiency" and "allocative efficiency." It discusses the distinction between "consumed fraction" and "nonconsumed fraction" in irrigation water use, and their implications for measures to improve water use efficiency and save water in irrigated agriculture (Perry, 2007), and the basin water accounting framework of Molden et al. (2003). It discusses the analytical procedure for measuring water productivity at plant, field, system, and basin scales, and mechanisms for monitoring the same at various scales.

In the third section, water productivity trends are presented. The initial discussion is on how the growing land and water demand in agriculture in the world in general and various regions in particular, has been driving changes in water productivity in agriculture both at the field level and at the regional level in different regions. The time period considered is 1961/63–2005/07. It then goes on to show how the demand for agricultural commodities is expected to increase by 2030 and 2050 in different regions of the world and discuss its imperatives for land and water use, crop yields, and water productivity in each region, with particular reference to the factors that would drive water productivity improvements.

The agricultural productivity growth in South Asian countries in the past, particularly since 1980, is then covered; particularly the factors that have contributed to the changes. The next section discusses the imperative of the projected growth in demand for agricultural commodities in South Asia for raising crop water productivity (both at the field level and regional level), and how the same can be achieved through technical, institutional, and policy measures. This is followed by a structured discussion on the constraints in improving water productivity of both rain-fed and irrigated agriculture in South Asian countries, with specific references.

In the last section, we will discuss a few of the recent attempts by water resource bureaucracies in South Asian countries (India, Pakistan, Afghanistan, Bangladesh, and Nepal) to enhance water use efficiency and water productivity in irrigated agriculture, to compare their intended outcomes with the possible actual outcomes. Following this, a conceptual framework for identifying the nature of interventions required for enhancing water productivity at plant, field, system, and basin-scale is presented. The framework classifies the river basins of South Asia into six categories for analyzing the WP improvement options on the basis of renewable water availability, how much of it is appropriated for meeting the existing demands in various sectors, what proportion of the current demands are met from the current supply potential and whether there are unmet demands in the basin which can be met from the untapped potential.

1.3 Contents of this book

This book is divided into 11 chapters. Following this introduction, Chapter 2 discusses a new methodology for assessing water and food security challenges of countries that distinguishes food security challenges from challenges posed by the shortage of water for meeting various direct consumption needs, which helps analyze water shortage for food production in relation to a quantum of arable land available for crop production and the weather parameters. It identifies countries that currently face a shortage in domestic agricultural outputs due to water shortage, and where water productivity improvements in agriculture should receive priority attention.

Chapter 3 explains the fundamental difference between the two concepts, viz., "water use efficiency" and "water productivity." It defines various dimensions of water productivity—physical productivity of water, combined physical and economic productivity and economic productivity of water, and a few others. The chapter also discusses how water productivity in agriculture needs to be seen in the light of total factor productivity (TFP), a measure of agricultural growth performance. The chapter also defines the terms "production function," "average water productivity," "marginal productivity of water," "technical efficiency" and "allocative efficiency of water used as input in crop production."

Chapter 4 discusses the methodologies for measurement/estimation of various water productivity functions at the plant, plot/field, system, and basin levels, which is a synthesis of published works of various scholars.

It also covers a detailed discussion on the indicators for measuring the performance of crops and agriculture with regard to changes/improvement in water productivity at various scales. It also provides a review of select studies that deal with the measurement/assessment of water productivity in agriculture at various scales, with a subsection on the impact of mulching on crop water productivity.

Chapter 5 reviews the past growth trends in agricultural water productivity globally and regionally, taking a 40-year time period. The regions considered for trend analysis are Sub Saharan Africa, Latin America, South Asia, South East Asia, East Asia, North and North East Africa, and the developed countries. The analysis is, however, confined to certain broad indicators of the resultant changes at the macro level such as rate of change in yield of major crops, rate of change in the value of agricultural outputs, and change in total water withdrawal for agriculture.

Chapter 6 reviews the situation in different regions of the world concerning the future possibilities for raising crop yields, expanding arable land, increasing irrigation potential, or intensifying cropping and improving water productivity for meeting agricultural growth challenges. It presents the scenarios of future growth in demand for agricultural commodities in different regions of the world due to projected increase in calorie intake and population growth, and supply of agricultural commodities owing to a projected increase in crop yields, expansion of arable land, and increase in irrigated area, based on a review of available studies on global institutions. Implications of these changes for agricultural water productivity in different regions are drawn.

In South Asia, expansion in private and public irrigation encouraged farmers who were earlier practicing low input agriculture, to intensify the use of external inputs and use of high yielding varieties and also shift to high-value cash crops. An overall increase in agricultural outputs due to yield improvements, intensive farming, and expansion of arable land, combined with diversification towards high-value commodities,[a] and development of market infrastructure, urbanization, and technological improvements resulted in growth in value of agricultural outputs in the region. In Chapter 7, we will systematically analyze the determinants of past growth in agricultural outputs of South Asia, and their implications for water productivity in agriculture.

---

[a] This diversification is driven by a growing demand resulting from rising per capita income.

Chapter 8 deals with strategies for enhancing agricultural productivity. The analysis deals with two regions, viz., South Asia and Sub Saharan Africa, covering the specific interventions for improving the utilization of water resources for crop production, technical efficiency improvements, other measures for improving water productivity in irrigated and rainfed crops, and the institutional and policy measures in the case of South Asia, and broad interventions in the case of Sub-Saharan Africa. The strategies for improving agricultural productivity and crop water productivity in the two regions are discussed with special reference to different agroecologies.

Chapter 9 discusses the constraints of introducing measures for improving water productivity in irrigated and rainfed areas of the two regions in a way that does not adversely affect economic benefits from farming. This means that any intervention to raise water productivity gain should lead to higher or same net return per unit of land while raising the returns per unit of water. The types of constraints included socioeconomic, financial, and social in the case of rainfed areas, and physical, institutional, policy-related, market-related, technological, and environmental in the case of irrigated areas.

In Chapter 10, we argue that to understand the need for and the opportunities and constraints of improving water productivity and to know where such measures would yield benefits, we need a framework to characterize the river basins on the basis of the following attributes: the renewable water available in the basin, the extent of its appropriation for meeting the existing demands, including agriculture, the proportion of the current demands that are met from the current supply potential and the unmet demands in the basin which can be met from further exploitation of the resources. We will first discuss how some of the measures implemented in the past to improve water use efficiency and water productivity in agriculture, based on poor understanding of the basin characteristics, turned out to be ineffective. In the next section, we will present an analytical framework that helps characterize the river basins and identify the water productivity improvement measures that would be effective at various scales, i.e., plant-level, plot-level, irrigation system level, and basin level.

In the concluding chapter (Chapter 11), we will summarize the findings from each chapter. We will also present the important messages for the global audience regarding the underlying concerns of, and objectives, and measures for improving water productivity in agriculture, and the outcomes that can be expected from such measures.

References

Ahmad, M. U. D., Masih, I., & Turral, H. (2004). Diagnostic analysis of spatial and temporal variations in crop water productivity: A field scale analysis of the rice-wheat cropping system of Punjab. *Journal of Applied Irrigation Science, 39*(1), 43–63.

Beddington, J. (2010). Food security: Contributions from science to a new and greener revolution. *Philosophical Transactions of the Royal Society B: Biological Sciences, 365*(1537), 61–71.

Falkenmark, M. (1989). The massive water scarcity now threatening Africa: Why isn't it being addressed? *Ambio*, 112–118.

Falkenmark, M. (1997). Meeting water requirements of an expanding world population. *Philosophical Transactions of the Royal Society of London. Series B: Biological Sciences, 352*(1356), 929–936.

Frederiksen, H. D., & Allen, R. G. (2011). A common basis for analysis, evaluation and comparison of offstream water uses. *Water International, 36*(3), 266–282.

Frederiksen, H. D., Allen, R. G., Burt, C. M., & Perry, C. (2012). Responses to Gleick et al., which was itself a response to Frederiksen and Allen. *Water International, 37*(2), 183–197.

Gleick, P. H., Christian-Smith, J., & Cooley, H. (2011). Water-use efficiency and productivity: Rethinking the basin approach. *Water International, 36*(7), 784–798.

Kumar, M. D., Bassi, N., Singh, O. P., Ganguli, A., & Chattopadhyay, S. (2017). *Assessing global water and food challenges: Time to rethink on the methods?*. Occasional Paper # 12 Hyderabad: Institute for Resource Analysis and Policy.

Kumar, M. D., & van Dam, J. C. (2009). Improving water productivity in agriculture in India: Beyond 'More Crop per Drop' (no. 612-2016-40570). In M. D. Kumar, U. A. Amarasinghe, B. R. Sharma, K. Trivedi, O. P. Singh, A. K. Sikka, & J. C. van Dam (Eds.), *Strategic analyses of the National River Linking Project (NRLP) of India series 5. Water productivity improvements in Indian agriculture: Potentials, constraints and prospects* (p. 163).

Kumar, M. D., & van Dam, J. C. (2013). Drivers of change in agricultural water productivity and its improvement at basin scale in developing economies. *Water International, 38*(3), 312–325.

Kummu, M., Guillaume, J. H., de Moel, H., Eisner, S., Flörke, M., Porkka, M., ... Ward, P. J. (2016). The world's road to water scarcity: Shortage and stress in the 20th century and pathways towards sustainability. *Scientific Reports, 6*, 38495.

Mancosu, N., Snyder, R. L., Kyriakakis, G., & Spano, D. (2015). Water scarcity and future challenges for food production. *Water, 7*(3), 975–992.

Molden, D., Murray-Rust, H., Sakthivadivel, R., & Makin, I. (2003). A water-productivity framework for understanding and action. In *Water productivity in agriculture: Limits and opportunities for improvement, (1). Comprehensive assessment of water management in agriculture.* UK: CABI Publishing in Association With International Water Management Institute.

Perry, C. (2007). Efficient irrigation; inefficient communication; flawed recommendations. *Irrigation and Drainage: The journal of the International Commission on Irrigation and Drainage, 56*(4), 367–378.

Raskin, I., Smith, R. D., & Salt, D. E. (1997). Phytoremediation of metals: Using plants to remove pollutants from the environment. *Current Opinion in Biotechnology, 8*(2), 221–226. https://doi.org/10.1016/s0958-1669(97)80106-1.

Seckler, D. (1996). *The new era of water resources management: From "dry" to "wet" water savings.* Colombo, Sri Lanka: International Irrigation Management Institute (IIMI), IWMI. iii, 17p. (IIMI Research Report 1). https://doi.org/10.3910/2009.003.

Van Dam, J. C., Singh, R., Bessembinder, J. J. E., Leffelaar, P. A., Bastiaanssen, W. G. M., Jhorar, R. K., ... Droogers, P. (2006). Assessing options to increase water productivity in irrigated river basins using remote sensing and modelling tools. *Water Resources Development, 22*(1), 115–133.

CHAPTER 2

Implications of a new methodology for assessing global water and food security challenges for agricultural water productivity

2.1 Introduction

Water is needed for growing food, feed, and fodder. It is also needed for human and animal consumption; for a myriad of functions—manufacturing processes, power generation, and finally, the environment. Globally, food, fodder, and feed production consume the largest amount of water currently and it would continue to be so for many decades to come as per projections (FAO, 2017). One major factor of production in the agriculture sector is the availability of land (Alexandratos & Bruinsma, 2012; FAO, 2011; Helfand & Levine, 2004; Kumar & Singh, 2005; Zhao, Hubacek, Feng, Sun, & Liu, 2019). Yet, the challenges of growing food, feed, and fodder, and challenges of meeting water needs of domestic, urban, industrial, and environmental sectors are all generally clubbed underwater management challenges, perhaps due to the crucial role water plays in the day-to-day life of beings as a life-saving and life-threatening resource. Hence, the global, national, and regional challenges of meeting future agricultural production needs and water supplies are viewed purely from a water resource perspective (Kumar, Bassi, & Singh, 2020).

Based on estimates of water demand for all the above-mentioned sectors in different regions and its comparison with renewable water supplies in those regions, global water scarcity maps were created. Such maps are used to arrive at inferences about the nature of water scarcity—whether physical or economic (see Seckler, Amarasinghe, Molden, de Silva, & Barker, 1998),

Current Directions in Water Scarcity Research, Volume 3
https://doi.org/10.1016/B978-0-323-91277-8.00001-0

and the degree of scarcity, whether in the category of "absolute water scarcity" or "water scarcity" or "water stress" (Falkenmark, 1989). The inferences are used to draw conclusions on whether a region or country would face a shortage of water for food production or economic growth or even basic survival needs (Kumar et al., 2020), and measures such as agricultural water productivity improvement are required. The best example is the widely used physical water scarcity index, which is used to indicate the sufficiency of water for economic production functions including agricultural production, and basic survival needs in a region, merely based on the per capita renewable water resources (see Falkenmark, 1989, 1997). This is a highly water-centric approach (Kumar et al., 2020; Zhao et al., 2019).

Such an approach had led to a skewed understanding of the nature of challenges the world would face in the future concerning water and food (Kumar et al., 2020; Kumar & Singh, 2005). For instance, estimates of renewable water supplies below the aggregate demand of water for meeting various future needs (of food production, water supplies for human, and livestock drinking and manufacturing) for any given region eventually led to the conclusion that the said region would face water scarcity problems (Kummu et al., 2016; Mancosu, Snyder, Kyriakakis, & Spano, 2015; Rijsberman, 2006; Seckler et al., 1998), and raising agricultural water productivity would be urgently needed there. An index that captures water use against water availability, called "criticality ratio" was derived by Raskin and others (Raskin, Smith, & Salt, 1997). As per this approach, a country is said to be "water scarce" if annual withdrawals are between 20% and 40% of annual supply, and severely water scarce if they exceed 40% (White, 2014). Such an approach does not factor in the differences in the quantum of water required to produce a kilogram of food-grain and other agricultural produce between different climatic settings, say between hot and arid tropics and cold and temperate regions (Mekonnen & Hoekstra, 2011). For instance, Mekonnen and Hoekstra (2011) showed that the average water footprint for cereal production ranged from 3388 m^3/ton in Africa to 1969 in Oceania to 1774 in Asia to 1294 in Americas to 1214 in Europe. Nevertheless, such figures are transposed with projected population figures to predict future water scarcity. They have also become the basis for "doomsday prophecies" (Kumar et al., 2020).

But assessing the food security challenge purely from a renewable water resource or a water use perspective provides a distorted view of the food security scenario of that country, by bringing in complacency among water-rich nations that they could be food-secure, and unwanted pessimism for water-scarce nations that they can't produce sufficient food unless urgent measures are taken to raise water productivity in agriculture. The point is that access to arable land is equally or even more important for food security and therefore should be integrated with other considerations in national food and water policymaking. In the same manner, assessing the water management challenges posed by nations purely from the point of view of renewable water availability and aggregate demands will be dangerous (Jia, Long, & Liu, 2017; Zhao et al., 2019). Access to water in the soil profile would be an important determinant of effective water availability for food production (Bodner, Nakhforoosh, & Kaul, 2015; Rockström et al., 2009), from which a major portion of the aggregate demand for water comes (see, for instance, De Fraiture & Wichelns, 2010; Kumar & Singh, 2005; Hanjra & Qureshi, 2010). Conversely, the large size of cultivated land in per capita terms increases the demand for water in the agriculture sector and can induce a water scarcity situation, even with moderately high per capita renewable water resources. This means that such regions would require measures for improving agricultural water productivity. Yet, food security challenges are often expressed in terms of water availability for irrigation, ignoring the contribution of land (Beddington, 2010; Kumar et al., 2020; Mancosu et al., 2015).

Such methodological drawbacks also lead to misleading conclusions about the magnitude of water scarcity and food insecurity problems in different countries and regions within countries (Kumar et al., 2020; White, 2014). Often, future food demand is translated into equivalent future water demand, and the same is compared against the utilizable water supplies (Kumar et al., 2020; Kummu et al., 2016; Mancosu et al., 2015). If the latter exceeds the former, suggestions are made to increase the utilization of the available water and improve the access to water (see, for instance, Rosegrant, Ringler, & Zhu, 2009; Namara et al., 2010; Mukherji, Shah, & Banerjee, 2012), completely ignoring the critical question, i.e., whether sufficient land to produce food using this water is available or not (Kumar et al., 2020).[a]

This new approach in analyzing food and water challenges, which integrates arable land as a key variable for estimating both the supply and demand

---

[a] Ideally, the following assessments are to be made: (a) the maximum amount of water that the available arable land in the region in question can absorb for irrigation under the highest possible intensity of land use for crop production; (b) the maximum production possible under the most intense land use; and, (c) the amount of food that has to be imported (given the likely future deficit) from other regions that have surplus land (Kumar et al., 2020).

of water for food and agricultural production, and production of food and other agricultural commodities, will force us to revisit the estimates of future food production requirements for regions or countries that have an excess amount of arable land. Obviously, in addition to meeting their own demand, such regions and countries also will have to produce a surplus for export to regions that are land scarce. The estimation of future water demand for agriculture in such regions should therefore include the quantum of surplus food for export to the neighboring food-insecure regions or countries that are the potential buyers of these agricultural commodities, even if the former is water-scarce.

This is quite contrary to the conventional wisdom of suggesting food import by such regions to reduce their water footprint (Allan, 1997; Hoekstra, 2003a, 2003b), which does not take into account the fact that water-rich regions often lack sufficient arable land to produce food even to meet their own needs, due to high population density. The new approach will also force us to revisit the estimates of future water requirements for various uses for regions that are land-scarce (Kumar et al., 2020). It will also automatically demand greater emphasis on raising water productivity and lowering water footprint in agricultural production in such regions.

Such an approach would give a new direction to the global debate on national and regional food and water policy. The reason is that it very clearly brings out the limitations that many water-rich countries and regions currently face or would face in securing sufficient food and agricultural commodities for meeting their domestic needs and therefore the type of interdependence that currently exists or is likely to develop between countries and regions within countries in the future for achieving water and food security, in the form of intercountry and interregional water transfer and intercountry and interregional food trade.

In this chapter, we attempt the following: (1) assessing the nature of challenges facing different countries, i.e., whether it is food security-related or water (supply and demand) management-related; (2) estimating the water footprint in agricultural production and diet of different countries; (3) analyzing the scope for improving food security and water supply situation in the countries that are confronting these problems of food insecurity and water shortage today, by examining the current level of utilization of water in the soil profile for agricultural production, blue water for water supplies and agricultural production, cultivation of cropland for overall agricultural production and use of pasture land for dairy production; and (4) assessing the need and scope for reducing the agricultural water footprint of individual countries by comparing their water footprint in diet and effective water withdrawal for production of agricultural commodities with global averages and also with effective renewable water availability in the respective countries. These assessments are used to identify countries and regions where water is actually a constraint for enhancing agricultural production and where water productivity improvement is necessary for enhancing agricultural production and mitigating food insecurity, and/or leaving more water for the environment.

2.2 Approach and methodology

The basic premise for the analysis and arguments presented in this chapter is that water abundance or water richness doesn't help a country to have large agricultural production to achieve food security and have a surplus for export. Instead, what is important is to have an adequate amount of cultivable land, supported by an adequate amount of utilizable water resources that can meet the crop water requirement of the available land, for the entire crop growing season. This means, there is an optimum level of water availability for a given amount of cultivable land for achieving the highest agricultural output (Kumar et al., 2020). If that much water is not available, countries should start looking at water productivity improvements in agriculture as an option to increase farm outputs. Conversely, countries with abundant water resources, but not having a sufficient amount of arable land required to use that water for agriculture production need not worry about crop water productivity improvement.

The renewable water availability exceeding that level barely helps in increasing crop production and only creates a situation of water surplus, and the surplus water can be made available for meeting domestic, livestock, drinking, and industrial production purposes.

To this effect, an index called "water adequacy index" (WAI) was derived, which indicates to what extent the current water availability is sufficient to meet the maximum crop water requirement of the available cultivable (arable) land for an assumed crop-growing period of 300 days. Its value is estimated by dividing the total renewable water availability by the multiple of the total cultivable land and average annual ET_0. However, the maximum value that can be assigned for the water adequacy ratio is 1.0 (i.e., one) and accordingly, the values are adjusted. These values indicate the water adequacy for likely future needs also when the cultivated area increases and becomes equal to the cultivable area.

$$WAI = (AWR/(CA \times ET_0 \times 300)) \qquad (2.1)$$

where ET_0 is the average daily evapotranspiration for the country, as a whole; CA is the cultivated area in ha. If the value of the coefficient, WAI is more than 1.0, it is taken as 1.0.

If the value of WAI is less than 1.0, it is modified by dividing it by the fraction of the cultivable land that is actually under cultivation. If the original water adequacy index is 0.50, and if only 2/3rd of the arable land is under cultivation currently, the modified WAI becomes 0.75. If the value of the "modified WAI" for any country exceeds 1.0, it has to be adjusted to 1.0 (one), which is the maximum value.

The modified WAI is multiplied by the "per capita cultivated area (ha)" to derive a new index called "water-land index." The composite index is to capture the agricultural production potential of a country. The computed value of water-land index ranges from 0.0 (Sao Tome and Principe) to 2.056 (Australia).

Milk production, an important part of food and nutritional security, is a water-intensive farming activity and is an important component of the farm economy for many regions around the world. In addition to the drinking water use, there is a significant amount of water indirectly used by the dairy animals in the form of embedded water in the (green and dry) fodder and feed, the indirect use being several times higher than drinking water consumption. The intensity of water use for dairy production depends on the climate which heavily influences the water consumption for producing the feed and fodder (Sultana, Uddin, Ridoutt, & Peters, 2014).

Since milk production is dependent also on biomass (grass) production along with by-products from cultivated land, in order to understand the effect of the same in milk production, a pasture land index was derived by multiplying the amount of pasture land per capita (ha) and a "coefficient," which reflects the quality of the pasture land in terms of primary productivity. The value of the "pasture land coefficient," which is used to simulate the variation in the quality of pasture land in terms of primary productivity, across countries, is decided on the basis of the rainfall and the reference evapotranspiration, and the overall climate (tropical or temperate). Up to 300 mm of rainfall, the coefficient considered is the same as that of the rainfall expressed in "meter." Beyond 300 mm of annual precipitation, the value of the coefficient considered is only a fraction of the rainfall, with its value increasing with a more uniform pattern of occurrence. The higher value for the coefficient is assigned in cases where the rainfall is high, and reference ET is low, and the climate is cold or temperate, and vice versa for low rainfall regions with high aridity, and tropical climate. The value of the pasture land coefficient chosen for the analysis varies from 0.05 for Saudi Arabia, UAE, Bahrain and Qatar to a maximum of "1.70" for New Zealand.

The sum of the multiple of modified water adequate ratio and the available cultivated land per capita (Mod. WAI × CULTIVATED LAND) and pasture land index is called "composite water-land-pasture land index."

The three indices, viz., modified water adequacy index, water-land index, and water-land-pasture index were used to assess the comparative situation of 153 countries around the world in terms of self-sufficiency to meet the future water needs, and the agricultural production potential including milk production potential, respectively. The computed values of the last two indices for the individual countries are used as independent variables in statistical models to explore how far they could predict the agricultural production and milk production potential respectively of the respective countries, thereby testing the robustness of the indices. Based on the nature of the relationship shown by the model, different typologies of countries were established in terms of their characteristics, defined by "water adequacy index" and "water-land index" and the consequent implications for water security and food security.

Irrigation potential utilization was estimated as a ratio of the total irrigation water withdrawal (m^3 per capita) against the total volume of water required to irrigate the cultivated land for nearly 300 days in a year. Subsequently, based on the current level of utilization of agricultural land for crop production, the extent of withdrawal of renewable water from the natural system, and the current level of utilization of irrigation potential, inferences are drawn with regard to the steps that these countries need to take to deal with future water scarcity and food insecurity challenges.

The values of effective renewable water resources are arrived at by adding up the AWR (actual annual renewable water resources) and the total amount of green water used in agriculture annually, by multiplying the gross cropped land and the effective rainfall or the amount of water in the soil profile for direct use by the crops—as a function of the total average annual rainfall of the country—as per the methodology used by Kumar and Singh (2005).

The water footprint in diet is estimated based on data on water embedded per unit of calorie in different types of food (cereals, milk, meat, vegetables, fruits, oil, etc.) and the average amount of calorie obtained from different types of food in different countries.

The analyses were carried out by extensively using global data sets on the following: (1) total population, geographical area, agricultural land, cultivated land, area under permanent crops, and area under pasture (ha) for the group of 172 countries, (fao.org/faostat/en/#data), milk production from different types of livestock and dairy animals, as available from FAO; (2) annual irrigation water withdrawal volumes by a group of 155 countries and irrigation potential utilization in percentage terms (as available from fao.org/faostat/en/#data); (3) estimates of green water use by a group of 155 countries available from Kumar and Singh (2005); and (4) data on virtual water export/import by a group of 155 countries available from Hoekstra and Chapagain (2006). The effective agricultural water withdrawal of the selected countries (131 nos.) was taken from Kumar and Singh (2005). Data on water embedded per unit of calorie in different types of food were obtained from Mekonnen and Hoekstra (2010); data on the average amount of calorie obtained from different types of food in different countries were obtained from FAO (undated) FAOSTAT database: http://www.fao.org/faostat/en/#data/CC which included data on food supply quantity (kg/capita/year) and calorie supply (kcal/capita/day), both for the year 2011.

2.3 The tenuous link between "renewable water" and food and agricultural production potential

If a region is having a sufficient amount of water resources in terms of the quantum of renewable water available, the automatic conclusion is that it can be food self-sufficient, if this water is utilized. The nexus between land and water, which define the food security and water management challenges, is not appreciated (Kumar et al., 2020; Kumar, Bassi, Narayanamoorthy, & Sivamohan, 2014). This nexus operates in two ways. First, even if a region doesn't have a sufficient amount of renewable water, the availability of a certain quantum of cultivable land (in the form of cultivated land, cultivable fallow, and pasture land) ensures the use of water in the soil profile, available directly from the precipitation, which can be used by the crop for production. Second, even if a region is highly water-rich (say owing to excessive precipitation), the absence of a sufficient amount of cultivable land can render the water unutilizable for crop production making the region food insecure.

We have analyzed the data on per capita renewable water availability in 150 countries and per capita cropped land (agricultural land).

Two samples paired T-test showed no relationship between per capita renewable water resources and per capita arable land. Many countries having a large amount of renewable water resources (in per capita terms) have poor access to agricultural land in per capita terms. On the other hand, many countries having a large amount of agricultural land (in per capita terms) have very limited water resources. There are very few countries, which are well endowed in terms of both renewable water resources and cultivable land. They are very unlikely to face any problems related to water and food, provided other factors of production such as labor, machinery, irrigation infrastructure, and (crop and irrigation) technologies are available. Now, we would examine how the positioning of a country vis-à-vis access to renewable water resources (including water in the soil profile) and agricultural land affect land and water utilization for crop production, food security scenario, and water available for the environment. The proportion of agricultural land under crop production would indicate land utilization, and the extent of water diversion from the hydrological system would indicate the degree of water utilization (Kumar et al., 2020).

If a country has a large amount of renewable water resources that are utilizable (say more than 20,000 m^3 per capita per year) and has a sufficient amount of agricultural land to feed its population, it would try and bring all that land cultivation to meet the food needs (countries like Cambodia, Vietnam and South Korea are examples), if other factors of production are available. Such countries would still have a lot of renewable water flowing into the natural sink, or neighboring countries (Kumar et al., 2020), given the limits on the amount of water that land can absorb in the form of evapotranspiration from agricultural crops and forest crops.

But, in such water-rich countries, if the amount of agricultural land available is more than what is required to produce agricultural commodities to meet the domestic needs, it is quite likely that a lot of that land would remain uncultivated as pasture land. Yet, the country might produce surplus food and other crops for export, like in the United States, Canada, and New Zealand. The result is that a lot of renewable water resources would remain untapped. Nevertheless, the situation of a good portion of the land lying uncultivated (and used as pasture) can happen also when the availability of renewable water resources is less than what is required to bring the entire land under cultivation, but more than what is required to feed the population. This is the case of Australia, which has very large per capita agricultural land (nearly 24 ha).

If the agricultural land available is too little, countries like Bangladesh and Indonesia would try and intensify land use, but might still have a food shortage, while a lot of blue water would be available as untapped runoff or groundwater. However, wealthy nations like Japan, which have too little arable land available for cultivation, would import food and leave the agricultural land fallow or convert it into pasture land (Kumar et al., 2020).

On the other hand, if a country has very little renewable water resources (less than what is required to meet the need of agricultural commodities), due to low precipitation and high aridity, but has a large amount of agricultural land, it would try and utilize all the water in the natural system for irrigated crop production, while also using the available soil moisture optimally through rainfed production by choosing the right season and right areas. In the process, it might also mine the fossil water (like many MINAR countries, Pakistan, Iran, and Iraq). In this case, there would still be a lot of lands left uncultivated in the form of pasture due to a shortage of water, though such land will offer low productivity.

If both renewable water resources and cultivable land are limited (say less than 1700 m^3 per capita per annum), the country would try and utilize all its renewable water resources along with water in the soil profile to increase the cropping intensity, by going for irrigated crops after harvesting rainy season crops in the same area. In such a situation, there will be much less water left in the natural system and there will be too little land left uncultivated. Such cases will be encountered in very arid regions, receiving very little precipitation. The western, northwestern, and parts of India (north Gujarat and Punjab) are the best example of this.

Therefore, in a nutshell, in countries having an excessively large amount of renewable water resources, even if the amount of land available for cultivation is large (which is not often the case though), the extent of utilization of water for agricultural production and needs of other sectors, would be relatively low, limited to what is required for producing sufficient food and other agricultural commodities for the domestic population and meeting domestic and manufacturing needs plus some water for producing surplus for exports. This is because of the potentially high costs involved in harnessing that additional water, with uncertain benefits from using that water. Whereas, in the case of countries having very limited renewable water resources, the degree of exploitation of available blue water would be very high (Kumar et al., 2020). This is evident from Fig. 2.1 (source: based on Kumar et al., 2020). In fact, Fig. 2.1 shows that the actual renewable water resources (sum of green water and blue water) determine the extent of water

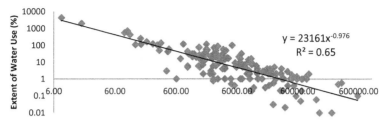

Fig. 2.1 Degree of blue water withdrawal vs effective per capita renewable water resources.

utilization to the extent of 65%. Hence, one of the major reasons why many countries are able to leave a lot of water in the hydrological system for nature is that they have plenty of renewable water resources in per capita terms.

On the other hand, in countries having an excessively large amount of cultivable land, even if utilizable water resources are available in plenty (which, again, is rarely the case), not all the land would be put under cultivation. The area that is put under cultivation, to a large extent, would be decided by the domestic demand for food grains and other agricultural commodities (oil seeds, vegetables, fruits, fiber, fodder, and fuel-wood) and only to a limited extent by the international demand for agricultural commodities, and the domestic policies for agricultural commodity trade. There are two reasons for the latter not becoming a major determining factor. There are many players in the global agricultural commodity trade, and the international agricultural commodity market can be highly sensitive to global agricultural outputs, which is determined by the domestic production surplus. Hence, as per capita cultivable land increases, the proportion of the agricultural land that is under cultivation will keep declining. The R2 value was 0.31 and the relationship was significant at a 95% confidence level.

Whereas there are countries in Africa, especially in eastern Africa, which are well endowed with ample amount of cultivable land and utilizable water resources, yet facing food shortages due to very little area under crop production. One reason is the poor infrastructure development for improving water supplies due to which the scale of irrigation is quite low (Kumar et al., 2020; Kumar, Saleth, Foster, Niranjan, & Sivamohan, 2016). There is a large amount of water flowing out of these countries into the downstream countries—from Ethiopia through the Blue Nile into Egypt. As the governance in these countries improves, the extent of water utilization would increase in the future, so will the chances for expansion in cultivated land, with the provision of irrigation facilities. The irrigation and (rural and urban) water supply situation are also likely to improve.

2.4 Water footprint in agricultural production and diet

The importance of managing the dietary regimes for future food security and the sustainable use of natural resources has been recognized (Falkenmark & Lannerstad, 2008; Foley et al., 2011; Mekonnen & Hoekstra, 2011; Rock-ström et al., 2009), and consumption of animal products that are at the higher end of the food chain is found to have large environmental impacts (Hoekstra & Chapagain, 2006; Steinfeld et al., 2006). There are studies on the impact of diets on water resources for countries (Liu & Savenije, 2008; Vanham, 2013) and regions (Renault & Wallender, 2000; Vanham & Bido-glio, 2013; Vanham, Mekonnen, & Hoekstra, 2013).

It is generally believed that one of the ways to mitigate water scarcity is by shifting to low water-intensive production systems, and following a dietary regime that comprises food items that are at the lower end of the food chain with a lower amount of embedded water, thereby reducing the overall water footprint in agricultural production, with a consequent reduction in agricul-tural water withdrawal (Hoekstra & Chapagain, 2006; Mekonnen & Hoek-stra, 2014; Vanham et al., 2013). Incidentally, the global database on agricultural water withdrawal by countries compiled by international agen-cies such as the FAO essentially consists of irrigation water withdrawal by these countries.

Though assessment of water footprint is separately available for blue and green water for several crops in different climatic regions in terms of vol-ume of water per ton of production (Mekonnen & Hoekstra, 2014), water withdrawal from soil profile is not considered in the international hydro-logical assessments. Obviously, the green water used by the crops grown in a temperate climate (per unit area) can be very significant (Hess, 2010) and far greater than that in semiarid and arid tropics. Omission of this compo-nent from water withdrawal estimates can send a misleading picture on water withdrawal for agriculture from the natural system (Hess, 2010), and an impression that the semiarid tropical countries use a disproportion-ately larger quantum of water for food production than the amount of food they produce.

Hence, the focus is always on countries, which draw large volumes of (blue) water for irrigation, especially those in the tropical climate, to reduce their agricultural water footprint through crop water productivity improvements (Mekonnen & Hoekstra, 2014). The green water used by countries, especially those in the temperate climate hasn't received much attention as they eventually draw a lower quantum of blue water for agriculture. Second, the extent to which several of the low irrigation water using (rich) countries in the temperate climate depend on imports of agricultural commodities, such as food grains, pulses, milk, oil, and fiber to meet their domestic demand and source of these imports are largely ignored to the extent that its impact on the agricultural water footprint of the food-exporting countries and the aggregate water footprint in the diet of the importing countries remains uninvestigated. Merely looking at the irrigation water withdrawal of the latter group or water footprint in dietary intake of the former group to address the issue of rising agricultural water footprint within the limited domain of the country of interest would only be misleading in terms of what options are available for countries to reduce their water footprint and its likely impacts.

In countries having a large amount of cropland (in per capita terms) and in countries having a temperate climate, the water withdrawal from soil profile (or green water use) constitutes a major proportion of the effective water withdrawal for agricultural production (Kumar & Singh, 2005). It is to be kept in mind that green water use is a major component of the hydrological balance, and change in green water use would affect the blue water flows (in the form of runoff and groundwater recharge) in the natural hydrological system. The ability to tap larger volumes of green water (by bringing large areas of the cultivable land under production and adjusting the cropping season to match the availability of precipitation for soil storage) reduces the irrigation requirement for crop production (Falkenmark, 2004). Therefore, it is important to look at this as a factor determining "effective water availability" as well as "effective water withdrawal" in agriculture.

Kumar and Singh (2005) provided estimates of effective water withdrawal in agriculture for 133 countries. It shows a wide range in the estimates among countries—from a meager 37 m^3 per capita for Malta to 18,965 m^3 per capita for Australia, i.e., nearly 1:500. The regression between green water use by 133 countries and the effective agricultural water withdrawal (sum of irrigation water withdrawal and green water use) shows a very high correlation ($R^2 = 0.97$), meaning that at the global scale, the effective water use for agriculture is determined by the extent of rainwater use—area under the crop, and the rainfall conditions.

Detailed assessment of water footprint at the national level had been done for several European countries, e.g. (Aldaya et al., 2010; Van Oel, Mekonnen, & Hoekstra, 2009), and countries outside Europe, e.g., (Bulsink, Hoekstra, & Booij, 2010). Detailed water footprint analyses on a global scale have been conducted for selected products, e.g., wheat (Mekonnen & Hoekstra, 2010) and rice (Chapagain & Hoekstra, 2011). We estimated the water footprint of diet for 155 countries based on the data on the average amount of calorie intake from different types of food and the water footprint in every unit calorie from these food types, for each country. Water footprint in the production of crops and livestock products is also a function of climatic factors and production practices (Mekonnen & Hoekstra, 2014). However, for the purpose of the study, average values of water footprint per unit of calorie intake for different types of food were considered. As per our estimates, the water footprint of food consumption ranges from the lowest of 480.8 m^3 per capita for Zambia to 2235 m^3 per capita for Argentina. In order to examine whether high agricultural water withdrawal is driven by domestic consumption behavior, we have run regression between the two for 119 countries for which data on both variables were available. The data are presented in Table 2.1.

The two-sample paired T-test showed no relationship between dietary water footprint and effective agricultural water withdrawal (effective Agricultural water footprint). Many of the countries which actually withdraw very little amounts of water from the natural system for crop production in per capita terms in the range of 240–400 m^3 per capita (viz., Germany, United Kingdom, Sweden, Switzerland, North Korea, South Korea, Israel, Jordan, Finland, France, and Japan), have a very high-water footprint in the diet. Even China, which has relatively low agricultural water withdrawal (670 m^3), maintains a diet of high-water footprint (1140 m^3 per capita). These countries import grains for direct consumption and as feed for dairy animals and livestock meant for meat production. They also import large quantities of vegetables, fruits, and flowers. Barring Israel, most of these are water-rich countries. But, the scope of increasing domestic production of agricultural commodities to reduce the import does not exist in these water-rich countries, due to land scarcity. At the same time, if a reduction in imports is accepted as a strategy to reduce dietary water footprint, it would help the food-exporting countries only if they experience resource depletion as well as water scarcity. For those, which are water-scarce (like China), changing food habits would help reduce domestic water withdrawals if import is continued.

Whereas, many countries effectively use excessively large quantities of water for agriculture such as Afghanistan, Angola, Argentina, Bolivia, Ecuador, Gabon, Guinea, Iran, Madagascar, Mozambique, Paraguay, Peru, Somalia, Sudan, Syria, Tajikistan, Uruguay, Uzbekistan, and Zambia, but have

Table 2.1 Effective agricultural water withdrawal and water footprint in diet of 119 countries.

S. no	Country name	Effective agricultural water withdrawal (M³/capita)	Water footprint in diet (M³/Capita)	S. no	Country name	Effective agricultural water withdrawal (M³/capita)	Water footprint in diet (M³/Capita)
1	Afghanistan	1624.9	660.2	61	Kuwait	142.5	1399.7
2	Albania	566.2	1404.9	62	Lao, PDR	879.2	671.1
3	Algeria	532.3	1107.3	63	Latvia	790.5	1122.6
4	Angola	4378.7	758.9	64	Lebanon	276.7	1350.8
5	Argentina	4213.5	2235.2	65	Liberia	871.4	572.2
6	Armenia	779.2	1301.4	66	Libya	930	1122.8
7	Australia	18,965.9	1732.1	67	Lithuania	746.6	1138.4
8	Austria	421.2	1468.3	68	Madagascar	2776.6	625.9
9	Bangladesh	663.1	553.2	69	Malawi	485.3	608.2
10	Barbados	146.5	1258.4	70	Malaysia	593.6	952.3
11	Belarus	774.8	1392.1	71	Mali	2286.2	951.7
12	Belgium	123.9	1357.7	72	Malta	37.4	1305.2
13	Belize	655.6	994.3	73	Mauritius	411.7	1024.2
14	Benin	483.4	655.2	74	Mexico	1457.5	1184
15	Bhutan	1246.4	1093.6	75	Moldova	622.8	933.2
16	Bolivia	4683.4	1106.6	76	Morocco	950.7	1074.8
17	Brazil	1771.7	1691.6	77	Mozambique	2809.1	556.9
18	Bulgaria	810.6	927.2	78	Nepal	653.9	742.1
19	Burkina Faso	734	697.7	79	Nicaragua	1632.1	832.7
20	Cambodia	791.7	599.7	80	Nigeria	610.2	710.5
21	Cameroon	679.4	802.2	81	Niger	670.9	790.2

Continued

Table 2.1 Effective agricultural water withdrawal and water footprint in diet of 119 countries.—cont'd

S. no	Country name	Effective agricultural water withdrawal (M³/capita)	Water footprint in diet (M³/Capita)	S. no	Country name	Effective agricultural water withdrawal (M³/capita)	Water footprint in diet (M³/Capita)
22	Canada	1555.4	1496.2	82	Norway	284.7	1359.4
23	C. Afric. Rep	1407.1	884.4	83	Pakistan	1305.9	867.6
24	China	669.6	1140.1	84	Panama	860.5	999
25	Colombia	1212.8	1223	85	Paraguay	4662.9	1004.6
26	Costa Rica	1151	1136.6	86	Peru	1891.8	786.9
27	Cote d'Ivoire	1354.3	740.3	87	Philippines	445.3	810.8
28	Cuba	1103.6	1170	88	Poland	392.2	1147.3
29	Denmark	474.8	1378.3	89	Portugal	1188.3	1412.9
30	Djibouti	1059.9	812.6	90	Romania	1083.8	1217.7
31	Dominican Republic	317.9	1130.7	91	Russia	836.1	1333.8
32	Ecuador	1774.9	1288.4	92	Rwanda	207.7	760.2
33	Egypt	873.4	1126.9	93	Senegal	804.8	680.8
34	El Salvador	397.1	829	94	Sierra Leone	628.5	564.2
35	Estonia	779.7	1205.9	95	Somalia	2984.9	718.7
36	Ethiopia	406.4	595.3	96	South Africa	1449.4	1210.5
37	Fiji	635.8	962.6	97	Spain	1179.6	1313.2
38	Finland	334	1532.7	98	Sudan	3372.8	955.4
39	France	446.7	1542.8	99	Suriname	1705.6	973.5
40	Gabon	4346.5	895.3	100	Sweden	296.4	1380.8
41	Gambia, The	427.8	694.2	101	Switzerland	228.3	1545.7
42	Georgia	988.8	867.4	102	Syria	1632.2	1123.6

#	Country			#	Country		
43	Germany	269.3	1312.5	103	Tajikistan	2156.8	691.2
44	Ghana	763.4	802.9	104	Tanzania	1270.4	653.3
45	Greece	1200.2	1511.9	105	Thailand	1689.8	786.4
46	Guatemala	551.3	809	106	Togo	844.6	549.3
47	Guinea-Bissau	1471.1	744.1	107	Trinidad and Tobago	118.3	1037
48	Haiti	323	653.8	108	Tunisia	716	1139
49	Honduras	579.3	920.6	109	Turkey	888.2	1349.2
50	Hungary	669.2	1023.9	110	Uganda	573.3	780.3
51	India	740.7	673.4	111	Ukraine	1016.8	1139.5
52	Indonesia	581.5	678.3	112	Un. Kingdom	289.9	1426.3
53	Iran	1535.9	1165	113	United States	1825.7	1662.7
54	Israel	256.2	1801.3	114	Uruguay	5407.5	1531
55	Italy	550	1639.8	115	Uzbekistan	2795.1	1372.2
56	Jamaica	204.7	978.5	116	Vietnam	737.1	940.1
57	Japan	477.5	867.1	117	Yemen	682.4	704.8
58	Jordan	235.5	1123.4	118	Zambia	3707.3	480.8
59	Kenya	706.7	827.7	119	Zimbabwe	1425.6	642.6
60	Korea, North	232.6	1159.5				

relatively low water footprint in their dietary intake—mostly less than 1000 m^3 per capita per annum. Many of them export agricultural commodities. Of these countries, the focus should be on those which are water-scarce, such as Afghanistan, Iran, Tajikistan, Uzbekistan, Somalia, Sudan, Syria, and Zambia. These countries will achieve much not by changing their diet, which has a low water footprint, but by reducing the export of agricultural commodities. However, this will be at the cost of their export economy. The fact that a lot of virtual water flows from water-scarce and land-rich countries to water-rich and land-scarce countries was also established by Kumar and Singh (2005).

Also, some countries have very high agricultural water withdrawal, and also maintain diets that have high water footprint, such as Australia, Argentina, Canada, and to an extent the United States. Among these, Australia and the United States (many parts of which experience severe water scarcity and droughts) can reduce their agricultural water withdrawal, provided the countries (especially in Europe) importing agricultural commodities—dairy and livestock products, vegetables, fresh and dry fruits, etc., from these countries change their consumption patterns. Also, the very countries exporting these agricultural commodities can change their consumption patterns.

2.5 Which countries have plenty of water for future crop production?

Kumar and Singh (2005) analyzed the determinants of global virtual water trade found that virtual water export from a country is a direct function of access to arable land, expressed in terms of gross cropped area, and explained this phenomenon using the concept of "effective agricultural water withdrawal," the sum of irrigation water withdrawal and water use from the soil profile, which increased linearly with per capita gross cropped land. They further developed a framework according to which the country which has a large amount of agricultural land with sufficient amount of water to bring in under crop production would have the highest advantage in terms of producing agricultural surplus and export. However, a country that has the same amount of water, but with very little agricultural land, will not enjoy the same advantage. On the contrary, a country with less amount of renewable (blue) water, with the same amount of agricultural land as that of the first country, will be relatively better off in terms of the agricultural production scenario. This essentially means, beyond a point, having a large amount of water disproportionately higher than the amount of cultivated land, doesn't help in increasing agricultural production.

We developed a composite water-land index, which captures two important dimensions of the future agricultural production potential of a country, viz., adequacy of water to bring the available cultivated land under intensive irrigated production (water adequacy index) enough for 300 days in a year, and the total amount of cultivated land in per capita terms, by multiplying them. The water adequacy index was estimated as a ratio of the total amount of renewable water resources and the amount of water required for irrigating the entire cultivated land for a period of 300 days. The maximum value of the water adequacy index is 1.0, wherein a computed value above "1.0" is treated as one. The value of the composite (water-land) index varies from 0.0 (for the Solomon Islands) to 2.056 for Australia. The regression run between the water-land index and the per capita virtual water export showed a linear relationship with an R^2 value of 0.55 (Kumar et al., 2020). The results are presented in Fig. 2.2 (source: based on Kumar et al., 2020). The relationship is significant at a 95% confidence level. An increase in the value of the water-land index meant greater agricultural surplus after domestic consumption, and a reduced value of the index meant a production deficit to meet the domestic consumption needs. This strong relation means that the water-land index is robust enough to capture the agricultural production potential of a country.

There are 22 countries from around the world having water resources (having a water adequacy index of 1.0), which can be used to expand the cropped land by tapping the unutilized water resources for irrigation, thereby boosting agricultural production. They are Argentina, Belie, Benin, Bolivia, Brazil, Myanmar, Canada, Finland, France, Guatemala, Hungary, Lithuania, Nicaragua, Paraguay, Romania, Sweden, Thailand, USA, and Uruguay (Table 2.2). They are not doing so because of several reasons,

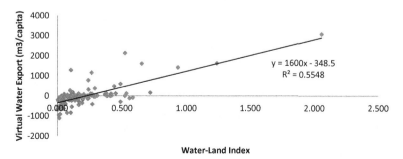

Fig. 2.2 Influence of water-land index on virtual water export of countries.

Table 2.2 Utilization of irrigation water and effective agricultural water withdrawal in countries having adequate amount of both arable land and water resources.

Name of country	Effective renewable water resource per capita (m³)	Effective agricultural water withdrawal (m³ per capita)	Water footprint per capita (m³)	Actual irrigation water use (m³ per capita)	Irrigation potential utilized (for water rich countries)	Per capita cultivated land (ha)	Virtual water export (m³ per capita)
Argentina	24,630	4213.5	2235.2	588.3	6.54	0.934	1418.15
Belize	71,760	655.6	994.3	0.9	0.04	0.238	312.31
Benin	4280	483.4	655.2	31.1	0.90	0.264	92.68
Bolivia	73,950	4683.4	1106.6	142.5	3.54	0.378	118.04
Brazil	47,150	1771.7	1691.6	217.8	6.39	0.365	71.43
Myanmar	2530	921.9		692.5	35.07	0.203	72.47
Burundi	2530	357.2		28.5	3.12	0.098	−0.69
Canada	92,852.5	1555.4	1496.2	137.0	3.58	1.239	1623.41
Finland	21,420	334.0	1532.7	12.8	0.82	0.417	90.43
France	3782.5	446.7	1542.8	66.8	3.99	0.288	376.35
Guatemala	9220	551.3	809.0	145.1	11.98	0.104	1283.04
Hungary	11,022.5	669.2	1023.9	239.7	9.58	0.439	475.59
Lithuania	8022.5	746.6	1138.4	5.7	0.16	0.719	162.61
Nicaragua	36,580	1632.1	832.7	218.6	7.15	0.255	102.47
Paraguay	60,440	4662.9	1004.6	65.3	0.86	0.653	1605.82
Romania	10,110	1083.8	1217.7	587.9	25.15	0.412	97.93
Russia	32,790	836.1	1333.8	93.3	3.14	0.833	−70.40
Sweden	19,857.5	296.4	1380.8	29.3	2.32	0.276	31.50
Thailand	6900	1689.8	786.4	1372.9	52.61	0.236	770.86
United States	11,485	1825.7	1662.7	712.0	23.61	0.488	589.85
Uruguay	44,990	5407.5	1531.0	914.7	17.90	0.522	2128.01

including marginal quality of the land available for future harvest, high cost of production, lack of domestic demand for the product, and therefore the need to tap the additional export market, etc. These are countries that have an excessively large amount of renewable (blue) water in per capita terms and a smaller fraction of their agricultural land under crop cultivation, leaving the remaining land under permanent crops and pastures. Essentially, these are also most likely to be countries that have a large amount of agricultural land in per capita terms (ranging from 0.49 for the United States to 1.24 for Canada), if we go by the earlier analysis. Their water-land index values range from 1.239 for Canada to 0.104 for Guatemala. So, merely by bringing a small portion of this agricultural land under crop cultivation, they are able to meet their needs for exports as well as domestic consumption. All of them are virtual water exporting countries (Kumar et al., 2020).

This is in contrast to countries like India, which have utilized nearly 90% of the agricultural land for (temporary) crop cultivation, given the very small amount of agricultural land per capita, resulting in excessively high anthropogenic pressure on land. Future growth in population in these countries is likely to result in an expansion in cultivated land. However, this is unlikely to happen in all cases, as population growth is either nil or negative in many of these countries that are developed.

Now there are 39 countries that do not have a sufficient amount of renewable water resources that can be used to bring all the cultivated land under irrigated production. Their per capita cultivated land area is in the range of 0.003 (Kuwait) and 2.069 (Australia). Their water adequacy index is less than one (Table 2.3). This means, even if all the utilizable water is harnessed and diverted for irrigation, it won't be sufficient to meet the irrigation demand of the entire cultivated area (not cropland). The low value of the water adequacy index is also a reflection of the good amount of cultivated land and high irrigation water demand. It is important to know how many of these countries are producing a sufficient amount of food, and if not, whether there is a possibility of increasing agricultural production in these countries (Kumar et al., 2020).

For this, we looked at the current level of utilization of irrigation water in these countries. Yet, among this, 12 countries use a small fraction of their irrigation potential. They are Australia, Denmark, and Burkina Faso, Sudan, Ukraine, Cuba and Moldova, Nigeria, Malawi, Uganda, Zimbabwe, and Afghanistan. Even with a small volume of blue water, they are able to produce surplus agricultural outputs (cereals in the case of Australia and dairy products in the case of Denmark) by tapping soil water, as they are endowed

Table 2.3 Utilization of irrigation water and effective agricultural water withdrawal in countries having limited water resources, but adequate amount of cultivated land.

Name of country	Effective renewable water resource per capita (m³)	Effective agricultural water withdrawal (m³ per capita)	Modified water adequacy index	Actual irrigation water use (m³ per capita)	Irrigation potential utilized (%)	Per capita cultivated land (ha)	Virtual water export (m³ per capita)
Afghanistan	3385	1624.9	0.859	886.5	42.5	0.267	8.75
Algeria	848	532.3	0.145	131.5	37.4	0.196	−349.88
Australia	42,822.5	18,965.9	0.982	951.3	4.8	2.094	3079.16
Barbados	376	146.5	0.679	75.1	31.7	0.036	−655.75
Bulgaria	3327.5	810.6	0.944	239.9	11.0	0.443	6.72
Burkina Faso	1605	734.0	0.147	62.7	8.4	0.356	81.45
China	2530	669.6	1.000	340.9		0.319	−15.52
Cuba	4080	1103.6	0.960	505.8	18.8	0.448	3.26
Denmark	1517.5	474.8	0.587	101.5	11.3	0.160	1148.46
Eritrea	2410	1009.1	0.574	72.7	6.2	0.170	−18.98
Ethiopia	2055	406.4	0.826	39.3	2.9	0.129	−5.22
India	1340	740.7	0.752	559.6	64.8	0.233	34.53
Iran	2515	1535.9	0.702	1055.3	67.0	0.039	−105.10
Israel	315	256.2	0.518	209.8	104.9	0.026	−915.09
Jordan	241	235.5	0.422	160.2	125.2	0.133	−956.71
Kenya	1605	706.7	0.560	34.4	4.6	0.003	−27.55
Kuwait	35	142.5	0.113	119.4	1866.3	0.288	−682.37
Libya	244	930.0	0.018	824.9	972.7	0.233	−258.60
Malawi	1810	485.3	0.542	80.2	7.2		74.70

Mali	9110	2286.2	0.810	648.8	10.9	0.474	−8.14
Malta	145	37.4	0.697	25.8	24.8	0.020	−1109.39
Moldova	3232.5	622.8	0.881	177.1	8.1	0.512	120.92
Morocco	1505	950.7	0.339	406.5	54.6	0.244	−202.72
Niger	3500	670.9	0.123	44.5	2.1	0.210	−43.52
Nigeria	2820	610.2	0.782	198.5	11.0	0.944	32.03
Pakistan	1600	1305.9	0.909	1206.0	106.2	0.121	−0.61
Qatar	131	410.3	0.738	372.5	541.5	0.006	−540.70
Rwanda	810	207.7	0.685	3.6	0.7	0.104	−11.82
Senegal	4470	804.8	0.651	154.1	5.1	0.251	−284.37
Somalia	3985	2984.9	0.664	386.8	35.0	0.114	−32.77
South Africa	2932.5	1449.4	0.377	264.5	29.8	0.231	−127.19
Sudan	4105	3372.8	0.121	1181.3	78.5	0.857	51.17
Tanzania	3640	1270.4	0.859	56.2	2.9	0.273	−26.83
Togo	3760	844.6	0.611	18.2	0.8	0.380	−145.52
Tunisia	965	716.0	0.166	236.0	64.1	0.264	−424.86
Uganda	3040	573.3	0.936	5.6	0.3	0.196	6.47
Ukraine	3560	1016.8	0.753	394.6	17.0	0.710	121.36
Yemen	733	682.4	0.283	370.5	233.9	0.050	−92.02
Zimbabwe	2810	1425.6	0.378	180.9	14.6	0.316	40.62

with a large amount of agricultural land per capita terms. If they expand irrigation, they would be able to increase their production manifold. However, their ability to trade these agricultural commodities globally depends largely on their cost of production and international market trends. All other countries are virtual water importing countries.

Among these 39 countries, there are seven countries whose water use far exceeds the renewable water availability. In fact, most of them are mining fossil groundwater. They are Jordan, Kuwait, Libya, Israel, Qatar, Yemen, and Pakistan. Yet, they are heavily dependent on virtual water imports. One of the reasons is that they have very little arable land that is productive and the rainfall is also too low. We can therefore consider the remaining 20 countries, where irrigation expansion through infrastructure development is possible to boost agricultural production. They are Algeria, Armenia, Barbados, Bulgaria, Cuba, Eritrea, Ethiopia, Iran, India, Kenya, Mali, Malta, Morocco, Niger, Rwanda, Senegal, South Africa, Tanzania, Togo, and Tunisia (Kumar et al., 2020).

Now, there is a third category of countries, 69 in total, which have a "water adequacy ratio" higher than one, and therefore considered as one (Table 2.4). In some cases, it is because they have very little cultivated land, and not because they have very large amounts of renewable water resources. As a matter of fact, out of the 69 countries, only four countries have more than 0.5 ha of cultivated land per capita and 29 countries have less than 0.10 ha of cultivated land per capita. Due to severe water shortages, some of them bring a small portion of their agricultural land under the cultivation of temporary crops (Bahrain, Oman, etc.). Nevertheless, many others have a large amount of renewable water, with very limited agricultural land and cultivated land (For instance, Bangladesh, Jamaica, El Salvador, Dominican Republic, etc.). But their distinct difference from countries such as the United States, Argentina, Brazil, and Canada (which have both large amounts of cultivated land and water resources) is that they have a disproportionately lower amount of arable land than water resources. This factor reduces their water-land index values. As a result, they are virtual water importing countries. Their irrigation potential utilization, estimated as a ratio of the total amount of irrigation water withdrawal (m^3 per capita) against the total amount of water required to irrigate the cultivated land for nearly 300 days in a year and expressed in percentage terms, ranges from as low as 0.1 (per cent) to 451 (per cent).

As evident from these, there are a few countries, which actually divert more water than what is required by the reported cultivated land. They are

Table 2.4 Utilization of irrigation potential and effective agricultural water withdrawal in countries having limited cultivated land, but having sufficient water to irrigate.

	Effective renewable water resource per capita (m³)	Effective agricultural water withdrawal (m³ per capita)	Water footprint per capita (m³)	Actual irrigation water use (m³ per capita)	Reference evapotranspiration (mm)	Virtual water export (m³ per capita)	Per capita cultivated land (ha)	Irrigation potential utilized (for water rich countries)
Albania	13,420	566.2	1404.9	312.9	2.4	−102.79	0.197	22.41
Angola	14,880	4378.7	758.9	16.0	3.9	−31.14	0.235	0.59
Armenia	3772.5	779.2	1301.4	510.7	2.80	−81.58	0.152	18.50
Bahrain	160	259.4	553.2	254.9	6.8	−640.82	0.003	438.48
Bangladesh	8210	663.1	1392.1	592.6	3.0	−43.81	0.050	131.91
Belarus	6587.5	774.8	1357.7	83.7	1.5	−89.17	0.584	3.10
Belgium	1882.5	123.9	1093.6	10.8	1.7	−846.22	0.075	2.77
Bhutan	41,650	1246.4		511.4	2.2	−34.01	0.132	59.47
Burundi	2530	357.2		28.5	3.1	−0.69	0.098	3.12
Cambodia	33,360	791.7	599.7	340.3	3.4	−11.08	0.274	12.09
Cameroon	18,150	679.4	802.2	50.2	3.4	−0.12	0.293	1.70
Central African Rep	38,320	1407.1	884.4	0.3	3.7	0.05	0.407	0.01
China	2530	669.6	1140.1	340.9	2.1	−15.52	0.076	72.41
Colombia	48,590	1212.8	1223.0	118.4	3.2	−161.34	0.033	37.16
Congo, Dem. Rep.	23,790			0.1	3.3	−3.69	1.663	0.00
Costa Rica	27,280	1151.0	1136.6	383.2	3.8	−221.19	0.053	63.11
Cote d'Ivoire	6110	1354.3	740.3	38.5	3.4	−49.45	0.149	2.53
Djibouti	1470	1059.9	812.6	11.3	5.1	−189.41	0.003	27.02

Continued

Table 2.4 Utilization of irrigation potential and effective agricultural water withdrawal in countries having limited cultivated land, but having sufficient water to irrigate.—cont'd

	Effective renewable water resource per capita (m³)	Effective agricultural water withdrawal (m³ per capita)	Water footprint per capita (m³)	Actual irrigation water use (m³ per capita)	Reference evapotranspiration (mm)	Virtual water export (m³ per capita)	Per capita cultivated land (ha)	Irrigation potential utilized (for water rich countries)
Dominican Republic	2790	317.9	1130.7	5.8	3.9	−6.06	0.079	0.63
Ecuador	32,960	1774.9	1288.4	1124.9	2.8	47.04	0.077	172.42
Egypt	835	873.4	1126.9	857.7	4.5	−293.22	0.035	180.56
El Salvador	4100	397.1	829.0	123.5	4.2	−236.65	0.110	8.93
Equatorial Guinea	52,030	752.7		2.3	2.7	−22.69	0.238	0.12
Estonia	10,562.5	779.7	1205.9	5.8	1.4	−302.22	0.488	0.28
Fiji	34,290	635.8	962.6	62.3	3.2	−364.17	0.190	3.44
Gabon	125,710	4346.5	895.3	41.7	2.6	−192.46	0.194	2.76
Gambia, The	5882.5	427.8	694.2	15.8	5.3	−128.07	0.239	0.42
Georgia	13,150	988.8	867.4	410.6	2.2	−52.29	0.092	66.59
Germany	2027.5	269.3	1312.5	113.4	1.7	−159.91	0.143	15.51
Ghana	3240	763.4	802.9	13.2	4.1	−29.04	0.190	0.56
Greece	7472.5	1200.2	1511.9	593.1	3.0	−18.72	0.228	28.76
Guinea	21,860	1868.1		85.1	4.6	−4.94	1.241	0.49
Guinea-Bissau	27,620	1471.1	744.1	187.6	4.0	−8.12	0.030	53.02
Haiti	1870	323.0	653.8	119.2	3.5	−55.70	0.106	10.75
Honduras	14,000	579.3	920.6	110.3	3.9	−87.38	0.131	7.20
Indonesia	13,000	581.5	678.3	365.2	3.2	−111.79	0.097	39.01

Italy	3610	550.0	1639.8	347.2	2.4	−544.92	0.110	44.23
Jamaica	3720	204.7	978.5	7.7	3.6	−174.22	0.044	1.64
Japan	3440	477.5	867.1	436.2	2.0	−645.69	0.034	212.71
Korea, North	1520	232.6	1159.5	190.4	2.2	−30.89	0.048	61.29
Korea, South	3580	352.7		224.0	2.4	−632.08	0.061	50.74
Lao, PDR	58,030	879.2	671.1	523.3	3.1	−19.75	0.219	26.06
Latvia	16,290	790.5	1122.6	16.6	1.5	−118.93	0.558	0.66
Lebanon	1280	276.7	1350.8	215.6	3.5	−446.11	0.027	75.74
Liberia	67,380	871.4	572.2	19.7	3.4	−27.08	0.123	1.60
Madagascar	20,800	2776.6	625.9	950.3	3.4	−9.57	0.161	57.36
Malaysia	23,700	593.6	952.3	246.4	3.0	−485.41	0.033	82.36
Mauritius	2340	411.7	1024.2	315.4	2.9	−407.10	0.062	57.89
Mexico	5275	1457.5	1184.0	624.5	3.7	−122.56	0.192	28.96
Mozambique	14,100	2809.1	556.9	31.7	3.8	−16.80	0.205	1.37
Nepal	8440	653.9	742.1	436.3	2.2	−1.26	0.080	82.16
Norway	84,180	284.7	1359.4	51.6	1.1	−450.52	0.163	9.28
Oman	490	661.1		523.3	5.2	−650.29	0.007	451.17
Panama	47,380	860.5	999.0	81.9	3.4	−37.26	0.145	5.62
Papua New Guinea	137,460	205.9		0.2	3.3	−99.87	0.043	0.05
Peru	95,170	1891.8	786.9	650.8	3.0	−230.30	0.134	54.70
Philippines	6080	445.3	810.8	284.5	3.2	−50.82	0.057	51.57
Poland	1960	392.2	1147.3	34.9	1.6	−42.69	0.290	2.59
Portugal	7195	1188.3	1412.9	878.5	2.5	−838.25	0.103	112.98
Spain	3355	1179.6	1313.2	615.0	2.9	−419.98	0.268	26.13
Suriname	278,360	1705.6	973.5	1493.6	3.4	173.45	0.118	124.74

Continued

Table 2.4 Utilization of irrigation potential and effective agricultural water withdrawal in countries having limited cultivated land, but having sufficient water to irrigate.—cont'd

	Effective renewable water resource per capita (m³)	Effective agricultural water withdrawal (m³ per capita)	Water footprint per capita (m³)	Actual irrigation water use (m³ per capita)	Reference evapotranspiration (mm)	Virtual water export (m³ per capita)	Per capita cultivated land (ha)	Irrigation potential utilized (for water rich countries)
Switzerland	7690	228.3	1545.7	7.0	1.6	−300.32	0.051	2.96
Tajikistan	3185	2156.8	691.2	1785.4	2.7	5.52	0.109	203.21
Trinidad and Tobago	3050	118.3	1037.0	15.5	3.5	−553.03	0.019	7.94
Turkey	3460	888.2	1349.2	433.0	3.0	−48.31	0.281	17.11
United Kingdom	2750	289.9	1426.3	4.7	1.3	−50.12	0.097	1.22
Uzbekistan	2555	2795.1	1372.2	2228.8	3.4	−29.34	0.153	141.92
Venezuela, Bol. Rep. of	48,080	1080.5		167.5	3.8	−218.63	0.093	15.72
Zambia	13,210	3707.3	480.8	133.7	4.3	9.69	0.262	3.93

Bahrain, Bangladesh, Ecuador, Egypt, Japan, Oman, Suriname, Tajikistan, and Uzbekistan. This excessive diversion could be because of the following reasons: (1) the actual water diversion for irrigation could be much higher than the consumptive water use we have considered, by using the figures of average ET and (2) some countries could be using perennial crops, due to which the total ET requirement of crops would be higher than what we have considered, which is for only 300 days. As regards the first point, for crops like paddy, a large amount of water is applied to the field to inundate it, which is far in excess of the crop ET requirements, and in certain cases, a lot of water would be required in hyperarid climates, to meet the nonbeneficial consumptive uses (such as barren soil evaporation). Yet, they are dependent on food imports. Unless the crop yields increase substantially, there is no way these countries can become food self-sufficient.

As Table 2.4 shows, there are 60 countries, which are not fully utilizing the irrigation potential yet remain as net importers of agricultural commodities. Many of them are developing countries with limited economic power, and want to boost their agricultural production. These countries can invest more in water resources development for irrigation intensification and thus become self-sufficient in agriculture, depending on the cost of water resources development, and the productivity of the land which is under cultivation. However, cropping to cover 300 days out of 365 days in a year is not an easy task. In any case, many of the developed countries (such as Germany and Italy) in this list will not be in a position to expand or intensify irrigation, due to scarcity of labor, and the high cost of production of crops. Also, in many countries in northern Europe and North America, the land is covered under snow for several months in a year, reducing the ability to go for intensive crop production.

2.6 Linkage between agricultural land, cultivated land, and milk production

From the above discussion, it is clear that having plenty of renewable water doesn't guarantee large agricultural production. Nor would plenty of cultivable land guarantee large-scale agricultural production. In fact, some countries have a large amount of land, classified under agricultural land (for instance, Kazakhstan and Mongolia). But the land is of poor quality, and at best is suitable for grazing (in Mongolia, even the pasture land is of poor quality, capable of producing grasses only for 4 months). What matters is the amount of land that is cultivated.

Without having water to irrigate and nutrients to supply to the soils, such marginal lands won't be able to produce much biomass, due to poor crop yields. The interaction between renewable water resources and cultivated land and between precipitation and pasture lands to produce biomass outputs is through a complex web, with climate at the backdrop. The amount of water required to irrigate a unit area of cropland (for intensive cropping to cover the whole year) is determined by the effective precipitation and climate. The climate determines the ET and effective precipitation decides the amount of irrigation, for a given crop ET. If an adequate amount of water is available to irrigate the cropped land, the overall production would depend on the total amount of cropped land. If this is not available, the productivity per unit crop area would decline. If more water (than what is required to irrigate the crop) is available, it would not lead to increased production, but increase the availability of water for ecological uses. The climate influences farming to a great extent also by allowing or disallowing certain types of crops and dairy animals and their varieties in an area.

Pasture land availability determines the livestock and dairy outputs. Very importantly, in any country, the expansion in arable land (cropped land) is at the cost of pasture land. As seen earlier, several countries that have large per capita cultivable land, have a large amount of pasture land and vice versa. The productivity of pasture land is a function of climate and precipitation. Year-round precipitation would ensure high productivity of pasture land with grasses if the climate is moderate (Kumar et al., 2020).

The best production system that illustrates this complex web of interactions is dairy production. Dairy animals survive on grazing in pasture land, stock feeding (of grasses or other fodder crops either from pasture land or from cropland), or both. In order to test the hypothesis about this complex web of interactions, we have developed a composite index for dairy production, which has a water-land nexus index and pasture land index.

The value of the coefficient is determined on the basis of the total annual precipitation and the average daily evapotranspiration, for obtaining a composite index for milk production capability. The higher the rainfall and lower the ET_0 value, the higher will be the value of the coefficient. The highest value of the coefficient assumed is 1.7 (for New Zealand) and the lowest value was 0.05 for hyperarid and desert countries (UAE, Saudi Arabia, Kuwait, Bahrain, Qatar, Yemen, Egypt, Oman, Libya, Algeria) with very low rainfall (200 mm and below). A high value of the coefficient indicates better primary productivity of the pasture land to produce biomass.

However, the full effect of climate in deciding the milk production potential of a region could not be captured in the composite index. Obviously, the type of dairy animals which a particular region can support depends on the climate there. For instance, a cold climate is very suitable for high-yielding breeds of cows (which are found in Northern, Central, and Eastern Europe and the Americas, New Zealand). Buffaloes are adapted to both hot and arid, and hot and humid tropical climate, and are seen in countries such as India, Pakistan, China, Nepal, Bangladesh, Afghanistan, Iran, Iraq, and some of the countries in Africa and Latin America.

A regression was run between per capita milk production and the composite index of milk production capability. The value of the composite index ranged from a lowest of 0.002 to a higher of 2.50. The per capita milk production is based on the total milk yield from five different types of dairy animals, viz., cows (both crossbred and indigenous), buffaloes, goats, sheep, and camels (mainly in the Middle-East) for the year 2013. The regression produced an R^2 value of 0.64, which means the composite index explains milk production to the extent of 64%. The relationship was found to be significant at a 95% confidence level.

The relationship suggests that with an increase in the value of the composite index, the per capita milk production also increases linearly. However, two trend lines clearly emerge, one with a steep slope and the other with a mild slope. Most countries on the steep gradient line are developed countries, experiencing cold and temperate climates. Most of the countries falling on the mild gradient line are developing countries, under a hot tropical climate. This differential productivity is mainly due to the differences in climate, which changes the production potential of the land and production technologies, and dairy production practices.

The countries that fall on the steeper gradient line, which we call the "technical efficiency frontier," generally have higher milk production efficiency. This can be due to the following facts: (1) favorable climatic conditions (high humidity, low temperature, and higher incident solar energy) leading to the production of higher biomass with lower transpiration, resulting in higher crop/fodder/grass yields and physical productivity of water; (2) adoption of certain high yielding varieties of cows such as Holstein Friesian and Jersey, which are better suited to the cold and temperate climates (Kumar et al., 2020); and (3) high energy conversion ratios due to stock feeding of animals.

A closer look at the model shows that India, whose composite index is a mere 0.113, maintains a per capita milk production of 113 kg per annum, whereas the value predicted by the model is only 16.3 kg per capita per annum. Therefore, India is on the "technical efficiency frontier," with production 7 times higher than the potential predicted by the model. On the other hand, New Zealand with a composite index of 4.25 produces 4836 kg per year, which is just 50% higher than the value (3254 kg/year) predicted by the model. Therefore, in spite of having several disadvantages vis-à-vis the production environment, India's performance in the dairy sector is superior (Kumar et al., 2020).

2.7 Findings and conclusions

The chapter analyzed the food security and water management challenges facing the world by delinking the food security challenges from the challenges of managing water supplies. This is based on the premise that: assessing the food security scenario purely from a water resource perspective gives a distorted view of the food security challenges facing a country (FAO, 2011; Kumar et al., 2020; Kumar & Singh, 2005; Rockström et al., 2009); and assessing the water management challenges purely from the point of view of renewable water availability also can be misleading (Kumar et al., 2020; Kumar & Singh, 2005). The chapter also critically examined the validity of the arguments for countries to reduce their agricultural water footprints, based on data on their irrigation water withdrawals and water footprint in their dietary intake. It did so by taking into account the following: water in the soil profile as part of the total water withdrawal for agriculture, and many countries maintaining a high diet with high water content, actually import large amounts of food. These two sets of analyses led to the identification of the regions and countries that need to be focused on improving water productivity in agriculture and reducing water footprint in agricultural production and diet.

The composite "water-land index" developed for this study captures the adequacy of water to bring the entire cultivated land under irrigated production (water adequacy index, whose maximum value is 1.0), and the amount of cultivated land per capita (ha). Subsequently, a composite index, which adds up the water-land index and the amount of quality pasture land (pasture land per capita x pasture land coefficient) available for grazing and grass production, was derived to predict the maximum per capita milk production in different countries. Subsequently, we assessed "water adequacy"; estimated "water-land index"; and estimated the composite index of water, cultivated land, and pasture land to stimulate the milk production potential in per capita terms for 152 countries, using global data sets (Kumar et al., 2020).

Analysis shows that availability of large amounts of renewable water resources (in per capita terms) doesn't mean water adequacy for irrigating all the cultivated land (water adequacy ratio < 1.0) if per capita cultivated land is large, and the country could face severe water shortages as the situation in Australia illustrates (Barrett, 2019). However, this doesn't have implications for agricultural production, as the country produces a large surplus for export (Qureshi, Hanjra, & Ward, 2013). As the situation of 10 countries illustrates, the availability of large amounts of renewable water resources, enough for bringing the entire cultivated land under irrigated production (water adequacy ratio > 1.0) and its full utilization also doesn't guarantee food self-sufficiency unless it is complemented by sufficient amount of arable land. The dependence on food imports increases for those water-rich countries, with lower utilization of irrigation potential (Kumar et al., 2020).

In sum, the criteria for assessing the magnitude of food security and water management challenges have to factor in the role of agricultural land, particularly the cultivated land. When this is done: we have four different categories of countries emerging: (1) countries with a large amount of renewable water and cultivated land, having both water and food self-sufficiency; (2) countries having large amounts of renewable water resources, but also having a disproportionately larger amount of cultivated land resulting in low "water-adequacy," but "food surplus," though facing occasional water shortages; (3) countries having sufficient amount of cultivated land, but low water availability, and facing different degrees of water shortages and food self-insufficiency; and (4) countries having high values of "water adequacy" most of the time because of a large amount of renewable water and low per capita cultivated land, and sometimes because of a disproportionately lower amount of cultivated land than what the available water can bring under intensive production, combined with low water availability, but mostly dependent on food imports (Kumar et al., 2020).

For the first set of countries, neither land productivity nor water productivity is a serious concern from the point of view of food security and water management, though they may invest in improving both if that helps increase the profitability of farming. For the second set of countries, improving land productivity will not be a concern, but improving water productivity in economic terms may be of interest as it would help them spread the available water to bring larger areas under irrigated crop or grass production, thereby increasing the total agricultural output in value terms and net income.

The third set of countries should be seriously concerned about improving water productivity in agriculture in both physical and economic terms. While doing this, they should also try and limit the area under crop production so as to cut down on aggregate water use in agriculture, thereby being able to free a significant portion of the water currently used in agriculture for nonagricultural uses in the future. For the fourth category of countries, improving biomass output per unit of land should be the concern.

Around 60 countries are belonging to the fourth category having large unutilized irrigation potential that can be tapped to increase cropping intensity. They can reduce their food imports, though some of them (especially in Europe) may not do it due to constraints of labor scarcity, high production cost, and ecological factors. Water productivity improvement in agriculture need not be a concern for them, as it might attract a lot of investment, without much gain. The reason is increased water availability through water productivity improvement will not result in any marginal gains for the economy or environment as plenty of water is already available in the natural system. On the other hand, around 20 water-scarce countries are belonging to the third category (mostly in Sub-Saharan Africa), with poor water adequacy for crop production, but also having varying quantities of water still remaining unutilized and can be harnessed to increase irrigation. Investment in water resources development will improve their food security situation (Alexandratos & Bruinsma, 2012). While doing this, they also need to focus on water demand management through water productivity improvements in crops and dairy farming.

Analysis of agricultural water footprint and water footprint in the diet for 150 countries shows that there is hardly any relation between the two. Many countries, where people maintain a diet with very high water content, actually do not produce much of the agricultural commodities domestically and instead import. Reduction in dietary water footprint though the change in consumption pattern would help, only if the countries that export food to these countries experience resource depletion. Whereas in many countries, which leave a very high-water footprint in their agricultural production, the water footprint of the average diet of the population is quite low, and the surplus production is exported. Some of them, viz., Afghanistan, Iran, Tajikistan, Uzbekistan, Somalia, Sudan, Syria, and Zambia. Reducing water footprint in their agriculture can be only through growing crops in a smaller area and this will be at the cost of their economy and livelihoods. There are many countries that leave a high-water footprint both in their diet, as well as in their agricultural production. However, only a few of them (Australia and the United States) experience water scarcity. To reduce the water footprint in agriculture, they need to change their food habits and also reduce agricultural exports. They can also improve water productivity in major cereals such as wheat, maize and rice, and dairy and meat production. But, for the latter to happen, the countries which import food from them need to change their consumption pattern.

To conclude, historically, analysis of water scarcity problems and food security challenges confronting nations around the world treated them as a single challenge, using simplistic considerations of renewable water availability and extent of withdrawal of renewable (blue) water from the natural system (Falkenmark, 1997; Kumar et al., 2020; Kummu et al., 2016; Raskin et al., 1997). Such skewed analyses ignore the role of access to agricultural land, particularly cultivated land (Kumar et al., 2020; Zhao et al., 2019) in determining the agricultural production potential of a country, while increasing the overall agricultural production potential itself. But increased access to cultivated land increases the access to water in the soil profile for crop production thereby reducing the demand for blue water in agriculture (Kumar et al., 2020; Kumar & Singh, 2005). Since the concerns about the need to raise water productivity in agriculture are linked to the overall demand for irrigation water (or blue water), such assessments are helpful in accurately choosing regions/countries for water productivity improvement measures.

The regions having a large amount of cultivated land per capita and sufficient water to cover the entire land under intensive production will have the comparative advantage for producing agricultural surplus for export (FAO, 2011; Zhao et al., 2019). This is followed by countries having high "per capita cultivated land," but a relatively low "water adequacy index" (39 in number). Countries that are very water-rich but having a very little amount of cultivated land in per capita terms (69 in number) will not face problems related to the water supply but might have to depend on food imports to meet domestic consumption needs. Nevertheless, there are many countries under the second and third category (20 out of the 39 nos. and 60 out of the 69 nos., respectively), which need to and can increase their production through increasing the utilization of available water for expanding irrigation and intensifying cropping.

Changes in consumption patterns would help countries having a high water footprint in their diet to reduce the water footprint since the countries from which the food is imported face water scarcity and resource depletion problems. Also, some water-scarce countries that leave a high water footprint in agriculture by producing a surplus and exporting, while maintaining a diet with a low water footprint, can reduce their water footprint by reducing agricultural commodity exports over a period of time, as their economic condition improves. Whereas some of the water–scarce countries, which maintain a diet with high water footprint (rich countries) and also produce surplus need to first focus on changing their consumption patterns and then reducing exports. For this, the consumption patterns of the food-importing countries need to change. If that doesn't happen, the major production base can be shifted to water-rich countries that are also land-rich, such as Argentina, Brazil, and Canada.

References

Aldaya, M. M., Garrido, A., Llamas, M. R., Varela-Ortega, C., Novo, P., & Casado, R. R. (2010). Water footprint and virtual water trade in Spain. In *Water policy in Spain* (pp. 49–59).

Alexandratos, N., & Bruinsma, J. (2012). World agriculture towards 2030/2050: The 2012 revision. In *ESA working paper # 12-03, Agricultural Economics Development Division, Food and Agriculture Organization of the United Nations, June 2012.*

Allan, J. A. (1997). 'Virtual water': A long-term solution for water short Middle Eastern economies? (pp. 24–29). London: School of Oriental and African Studies, University of London.

Barrett, J. (2019). Drought hit Australian towns prepare for 'unimaginable' water crisis. *Environment.*

Beddington, J. (2010). Food security: Contributions from science to a new and greener revolution. *Philosophical Transactions of the Royal Society B: Biological Sciences, 365*(1537), 61–71.

Bodner, G., Nakhforoosh, A., & Kaul, H. P. (2015). Management of crop water under drought: A review. *Agronomy for Sustainable Development, 35*(2), 401–442.

Bulsink, F., Hoekstra, A. Y., & Booij, M. J. (2010). The water footprint of Indonesian provinces related to the consumption of crop products. *Hydrology and Earth System Sciences, 14*(1), 119.

Chapagain, A. K., & Hoekstra, A. Y. (2011). The blue, green and grey water footprint of rice from production and consumption perspectives. *Ecological Economics, 70*(4), 749–758.

De Fraiture, C., & Wichelns, D. (2010). Satisfying future water demands for agriculture. *Agricultural Water Management, 97*(4), 502–511.

Falkenmark, M. (1989). The massive water scarcity now threatening Africa: Why isn't it being addressed? *Ambio,* 112–118.

Falkenmark, M. (1997). Meeting water requirements of an expanding world population. *Philosophical Transactions of the Royal Society of London. Series B: Biological Sciences, 352*(1356), 929–936.

Falkenmark, M. (2004). Towards integrated catchment management: Opening the paradigm locks between hydrology, ecology and policy-making. *International Journal of Water Resources Development, 20*(3), 275–281.

Falkenmark, M., & Lannerstad, M. (2008). Food security in water-short countries–Coping with carrying capacity overshoot. In L. Martinez-Cortina, A. Garrido, & E. Lopez-Gunn (Eds.), *Proceedings of the fourth Botin foundation water workshop: Re-thinking water and food security* (pp. 3–22). United States: CRC Press.

Foley, J. A., Ramankutty, N., Brauman, K. A., Cassidy, E. S., Gerber, J. S., Johnston, M., ... Balzer, C. (2011). Solutions for a cultivated planet. *Nature, 478*(7369), 337–342.

Food and Agriculture Organization of the United Nations. (2011). *The state of the World's land and water resources for food and agriculture-managing systems at risk.* New York: Earthscan and FAO.

Food and Agriculture Organization of the United Nations. (2017). *The future of food and agriculture–trends and challenges.* (Rome).

Hanjra, M. A., & Qureshi, M. E. (2010). Global water crisis and future food security in an era of climate change. *Food Policy, 35*(5), 365–377.

Helfand, S. M., & Levine, E. S. (2004). Farm size and the determinants of productive efficiency in the Brazilian Center-West. *Agricultural Economics, 31*(2–3), 241–249.

Hess, T. (2010). Estimating green water footprints in a temperate environment. *Water, 2*(3), 351–362. https://doi.org/10.3390/w2030351.

Hoekstra, A. Y. (2003a). Virtual water trade: A quantification of virtual water flows between nations in relation to international crop trade. In *Proceedings of the international expert meeting on virtual water trade 12, Delft, 2003* (pp. 25–47).

Hoekstra, A. Y. (2003b). Virtual water: An introduction. In A. Y. Hoekstra (Ed.), *Virtual water trade: Proceedings of the international expert meeting on virtual water trade. 12–13 December 2002.* The Netherlands: IHE Delft.

Hoekstra, A. Y., & Chapagain, A. K. (2006). Water footprints of nations: Water use by people as a function of their consumption pattern. In *Integrated assessment of water resources and global change* (pp. 35–48). Dordrecht: Springer.

Jia, S., Long, Q., & Liu, W. (2017). The fallacious strategy of virtual water trade. *International Journal of Water Resources Development, 33*(2), 340–347.

Kumar, M. D., Bassi, N., Narayanamoorthy, A., & Sivamohan, M. V. K. (Eds.). (2014). *The water, energy and food security nexus: Lessons from India for development* Routledge.

Kumar, M. D., Bassi, N., & Singh, O. P. (2020). Rethinking on the methodology for assessing global water and food challenges. *International Journal of Water Resources Development, 36*(2–3), 547–564. https://doi.org/10.1080/07900627.2019.1707071.

Kumar, M. D., Saleth, R. M., Foster, J. D., Niranjan, V., & Sivamohan, M. V. K. (2016). Water, human development, inclusive growth, and poverty alleviation: International perspectives. In M. D. Kumar, A. J. James, & Y. Kabir (Eds.), *Rural water systems for multiple uses and livelihood security* (pp. 3–22). Netherlands, UK and USA: Elsevier.

Kumar, M. D., & Singh, O. P. (2005). Virtual water in global food and water policy making: Is there a need for rethinking? *Water Resources Management, 19*(6), 759–789.

Kummu, M., Guillaume, J. H., de Moel, H., Eisner, S., Flörke, M., Porkka, M., ... Ward, P. J. (2016). The world's road to water scarcity: Shortage and stress in the 20th century and pathways towards sustainability. *Scientific Reports, 6*, 38495.

Liu, J., & Savenije, H. H. (2008). Food consumption patterns and their effect on water requirement in China. *Hydrology and Earth System Sciences Discussions, 12*(3), 887–898.

Mancosu, N., Snyder, R. L., Kyriakakis, G., & Spano, D. (2015). Water scarcity and future challenges for food production. *Water, 7*(3), 975–992.

Mekonnen, M. M., & Hoekstra, A. Y. (2010). *The green, blue and grey water footprint of crops and derived crops products.* Delft, The Netherlands: UNESCO-IHE Institute for Water Education.

Mekonnen, M. M., & Hoekstra, A. Y. (2011). The green, blue and grey water footprint of crops and derived crop products. *Hydrology and Earth System Sciences, 15*, 1577–1600.

Mekonnen, M. M., & Hoekstra, A. Y. (2014). Water footprint benchmarks for crop production: A first global assessment. *Ecological Indicators, 46*, 214–223.

Mukherji, A., Shah, T., & Banerjee, P. S. (2012). Kick-starting a second green revolution in Bengal. *Economic and Political Weekly*, 27–30.

Namara, R. E., Hanjra, M. A., Castillo, G. E., Ravnborg, H. M., Smith, L., & Van Koppen, B. (2010). Agricultural water management and poverty linkages. *Agricultural Water Management, 97*(4), 520–527.

Qureshi, M. E., Hanjra, M. A., & Ward, J. (2013). Impact of water scarcity in Australia on global food security in an era of climate change. *Food Policy, 38*, 136–145.

Raskin, I., Smith, R. D., & Salt, D. E. (1997). Phytoremediation of metals: Using plants to remove pollutants from the environment. *Current Opinion in Biotechnology, 8*(2), 221–226.

Renault, D., & Wallender, W. W. (2000). Nutritional water productivity and diets. *Agricultural Water Management, 45*(3), 275–296.

Rijsberman, F. R. (2006). Water scarcity: Fact or fiction? *Agricultural Water Management, 80*(1–3), 5–22.

Rockström, J., Falkenmark, M., Karlberg, L., Hoff, H., Rost, S., & Gerten, D. (2009). Future water availability for global food production: The potential of green water for increasing resilience to global change. *Water Resources Research, 45*(7).

Rosegrant, M. W., Ringler, C., & Zhu, T. (2009). Water for agriculture: Maintaining food security under growing scarcity. *Annual Review of Environment and Resources, 34.*

Seckler, D., Amarasinghe, U., Molden, D., de Silva, R., & Barker, R. (1998). *World water demand and supply, 1990 to 2025: Scenarios and issues* (pp. 68–110). Colombo: International Water Management Institute. Sri Lanka, Research Report 19.

Steinfeld, H., Gerber, P., Wassenaar, T. D., Castel, V., Rosales, M., Rosales, M., & de Haan, C. (2006). *Livestock's long shadow: Environmental issues and options.* Food & Agriculture Organization.

Sultana, M. N., Uddin, M. M., Ridoutt, B. G., & Peters, K. J. (2014). Comparison of water use in global milk production for different typical farms. *Agricultural Systems, 129*, 9–21.

Van Oel, P. R., Mekonnen, M. M., & Hoekstra, A. Y. (2009). The external water footprint of the Netherlands: Geographically-explicit quantification and impact assessment. *Ecological Economics, 69*(1), 82–92.

Vanham, D. (2013). The water footprint of Austria for different diets. *Water Science and Technology, 67*(4), 824–830.

Vanham, D., & Bidoglio, G. (2013). A review on the indicator water footprint for the EU28. *Ecological Indicators, 26*, 61–75.

Vanham, D., Mekonnen, M. M., & Hoekstra, A. Y. (2013). The water footprint of the EU for different diets. *Ecological Indicators, 32*, 1–8.

White, C. (2014). 28. Understanding water scarcity: Definitions and measurements. In R. Q. Grafton, P. Wyrwoll, C. White, & D. Allendes (Eds.), *Global water: Issues and insights.* Canberra, Australia: ANU Press, Australian National University.

Zhao, D., Hubacek, K., Feng, K., Sun, L., & Liu, J. (2019). Explaining virtual water trade: A spatial-temporal analysis of the comparative advantage of land, labor and water in China. *Water Research, 153*, 304–314.

CHAPTER 3

Conceptual issues in water use efficiency and water productivity

3.1 Introduction

Academic discourse on the two concepts, viz., "water use efficiency" and "water productivity" is quite old. While both concepts have evolved mainly in the context of agricultural water management, especially irrigation water management, they have different disciplinary origins and objectives. While the term "water use efficiency" is used to describe the effectiveness with which the water supplied in the field is used for plant growth or evapotranspiration, the term "water productivity" concerns the biomass outputs or economic value of those outputs produced from the use of the water applied or consumed by the crop. The latter is less concerned with what happens to the water that is consumed or applied, except when the measurement is done at different scales. Hence the drivers of change in water productivity are different from that of water use efficiency. Yet, both the terms are defined in many different ways. There is also a lack of conceptual clarity on the usefulness of each of the two terms or concepts in different contexts. In several instances, these terms are used interchangeably, with the result that different meanings are conveyed.

This chapter attempts to clarify the fundamental difference between the two concepts, viz., "water use efficiency" and "water productivity." It defines various dimensions of water productivity while touching upon the international debate on the conceptual issues and confusions surrounding the definitions of water productivity. The chapter also discusses how water productivity in agriculture needs to be understood in the light of total factor productivity (TFP). It also defines some of the other useful concepts used in analyzing the performance of the agricultural sector, such as production function, and average and marginal productivity of water, and technical and allocative efficiency in the use of water as an input for crop production.

Current Directions in Water Scarcity Research, Volume 3
https://doi.org/10.1016/B978-0-323-91277-8.00005-8

3.2 Water productivity vs water use efficiency

The concept of water productivity was introduced to complement existing measures of the performance of irrigation systems, mainly the classic irrigation and effective efficiency (Keller, Keller, & Seckler, 1996). Classic irrigation efficiency focuses on establishing the nature and extent of water losses and included storage efficiency, conveyance efficiency, distribution efficiency, and application efficiency (Gichuki, Cook, & Turral, 2006).

The classical definition of "irrigation efficiency" looks at the actual evapotranspiration (ET) against the total water diverted for crop production (Kijne, Randolph, & David, 2003). It fails to capture the water reuse aspect, especially while assessing the performance of irrigation systems at a system or basin scale. It ignores the beneficial use of the water which gets recaptured and reused in one part of the basin as a consequence of deep percolation and/or runoff losses that takes place elsewhere, and hence the "scale effect" (Ahmad, Masih, & Turral, 2004; Van Dam et al., 2006). Though it was used to analyze the "productive use" of water, it omitted economic values (Van Dam et al., 2006).

To address this problem, Keller et al. (1996) introduced the concept of effective efficiency, which takes into account the quantity of the water delivered from and returned to a basin's water supply. In an irrigation context, effective efficiency is the amount of beneficially used water divided by the amount of water used during the combined processes of conveying and applying that water.

Water productivity makes it possible to undertake a holistic and integrated performance assessment by including all types of water uses in a system and a wide variety of outputs; integrating measures of technical and allocative efficiency; incorporating multiple-use and sequential reuse as the water cascades through the basin; including multiple sources of water; and integrating nonwater factors (like agricultural inputs and market factors) that affect productivity (Gichuki et al., 2006. Water productivity (WP) is a partial-factor productivity that measures how the systems convert water into goods and services (Molden, Murray-Rust, Sakthivadivel, & Makin, 2003).

3.3 International debate on conceptual issues and confusions surrounding the definition of water productivity

There is rich debate internationally about almost every aspect of water productivity, starting from the usefulness of the concept of water productivity

(Zoebl, 2006) to the determinants of water productivity; and scale of measurement or unit of analysis of water productivity; and the methods of measurement of water productivity; and water-saving impacts of water productivity improvement measures in agriculture.

According to Zoebl (2006), rain or irrigation water being only one of the many inputs and growth conditions to ensure a certain output or yield, the term productivity is not appropriate and has to be reserved for genuine production factors such as labor, land, and capital. Besides, according to him, expressing this water productivity in output per unit water input also is not always meaningful. Unless the opportunity costs of forgone values are not taken into account, high water productivity values may not be a useful criterion for evaluating the performance of crop production systems. However, this is an overstatement given the fact that scholars had clarified that the water productivity measurements are measurements of partial productivity (Molden et al., 2003).

As regards the determinants of water productivity, after Kijne et al. (2003), there are three major crop water productivity parameters. They are physical productivity of water expressed in a kilogram of crop per cubic meter of water diverted or depleted (kg/m^3); net or gross present value of the crop produced per cubic meter of water (Rs/m^3) known either as economic efficiency of water use or combined physical and economic productivity of water; and net or gross present value of the crop produced against the value of the water diverted or depleted. Here, "the value of water" is the opportunity in the highest alternative use (Kijne et al., 2003). With three different water productivity parameters used, the determinants of water productivity can also change.

3.3.1 Physical productivity of water

For the physical productivity of water, the biomass output per unit volume of water is what matters. Whether the amount of water to be considered in the denominator should be the irrigation water applied, or the total of effective fall and irrigation or the total water consumed by the crop or total water depleted in the system would be dealt with in the section on water productivity measurements.

Physical productivity of water could be a good measure of the irrigation system performance when the same type of crop from two or more plots or fields are compared. Since biomass output per unit of water itself will be dependent on fertilizer input (Singh, van Dam, & Feddes, 2006: pp. 272), labor, climate, season, soil quality, etc., it is understood that these variables are controlled. Otherwise, such comparisons would be meaningful, only if the purpose is to understand the effect of any of the above-mentioned variables. For example, given the influence of climate on crop water productivity (Abdullaev & Molden, 2004; Zwart & Bastiaanssen, 2004), it is useful to compare (physical) water productivity of the same crop grown in two different agro climates receiving irrigation from the same type of source, if the idea is to study the effect of climate on crop WP. Nevertheless, here it is important to make sure that the soil type, fertilizer dosages, and variety selection are comparable. Here the differential water productivity could result from two factors. First, differential PET and rainfall, which changes the denominator of water productivity (irrigation as well as actual ET) (Abdullaev & Molden, 2004) and second, differences in yield which occur due to climatic factors, especially solar radiation and temperature (Aggarwal et al., 2008; Zwart & Bastiaanssen, 2004).

3.3.2 Combined physical and economic productivity of water

Since the yield (which is the production output) of the crop could be enhanced to a great extent through manipulation of inputs such as fertilizers, farm management involving labor, pesticides, use of high yielding varieties, which have economic implications in terms of input costs (Barker, Dawe, & Inocencio, 2003; Kumar & van Dam, 2013), but at the same time do not have any bearing on irrigation performance, assessing the performance of irrigation system, in situations where such manipulations are possible, will be complete only if water productivity, expressed in terms of economic output per unit volume of water ($/m^3) is also estimated along with physical productivity.

When the irrigation systems are different (say well irrigation from diesel-powered engines against gravity irrigation from canals or irrigation from electric wells or river lift), the cost of irrigation water in US$/m^3 would be different (Kumar, Singh, & Sivamohan, 2010). The system performance also will be different in terms of water control, timeliness, and reliability, with a direct impact on crop yield and water productivity (source: based on Meinzen-Dick, 1995; Kumar & Patel, 1995; Kumar et al., 2010). In such situations also, it is important to consider the net benefit against water input as one of the criteria for assessing the performance of the irrigation system. The reason is farmers are interested in maximizing the net benefit per unit of water or land, and the higher degree of reliability, water control, and timeliness, especially in the case of well irrigation comes with additional costs. Such broader criteria for assessing the performance of irrigated production system in economic terms become pertinent when the unit of analysis is the farm, as many different crops (Kumar, 2005), crops and livestock (Kumar, Scott, & Singh, 2011), with different values are involved.

3.3.3 Economic productivity of water

When performance assessment involves water from two different irrigation systems, the opportunity cost of depleting water could vary significantly as in the case of gravity irrigation from canals and well irrigation. Hence, a measure of output or net returns from a unit volume of water depleted would not be sufficient a criterion for assessing the performance. Instead, the net return per unit (opportunity) cost should also be examined.

There are two reasons for this. First, the scope for transferring water to alternative uses (having higher values) could be different between the groundwater system and surface water system. It is generally high for surface irrigation systems, as opportunities exist for transferring water to high-value uses such as municipal water supply and manufacturing.

Second, the return flows from gravity irrigation are likely to be different from that of well irrigation (Kumar & van Dam, 2013), and the cost of energy required for pumping out the return flows for reuse will also be different. The amount of water supplied in the field to get a unit rate of consumptive use is very high for surface irrigation. For instance, a water balance study in the Palleru river basin in Andhra Pradesh using the SWAT model showed that the consumptive use of water in irrigation in the command varies from 31.8% to 38.1%, indicating a return flow fraction as high as 68.2% to 61.9% (Gosain, Rao, Srinivasan, & Reddy, 2005). In a region, where agriculture is energy-intensive and energy is scarce, allowing excessive return flows in the form of recharge and runoff cannot be treated merely as a virtue, in spite of the benefits of groundwater recharge and the reuse value, as it would require precious energy to pump it for reuse. Estimating the "opportunity cost" of using water should therefore include the cost of producing the energy required for lifting the "return flow fraction" and the surplus-value product accrued from alternative uses of the "depleted water."

3.3.4 Other dimensions of water productivity

The above definitions of water productivity do not express the social benefit of agricultural water productivity explicitly (FAO, 2003). For instance, when speaking of food security, a useful approach is to look at the nutrient value of different crops, rather than merely looking at the quantum of cereals produced, given the differences in the nutritional value of different crops (Renaults & Wallender, 2000). But, if we look at livelihood security, which again is an important concern for developing countries (Booker & Trees, 2020; Kumar & van Dam, 2013), we need to look at the number of jobs per drop of water. There is no unique definition of water productivity and the value considered for the numerator might depend on the focus (FAO, 2003).

An analysis of farm-level water use in California shows that the greatest use of farm labor is concentrated in a relatively small number of cropping types. Expressed as a function of cumulative consumptive water use, three crop categories accounted for over half of farm labor use, but less than 20% of consumptive use. Further, it was found that 50% of the consumptive use of water was for crops that employ a full 95% of workers in direct farm production jobs (Booker & Trees, 2020). The analysis suggests with vegetables, nontree horticulture fruits, grapes, and almonds produced a large number of jobs per unit of water, crops such as alfalfa, cotton, onion, garlic, tomatoes, rice, sugar-beets, and other field crops that accounted for a larger share of the water consumption in the region did not produce many jobs (source: based on fig. 1 in Booker & Trees, 2020). In regions where high rural unemployment is as big a concern as water scarcity, apart from the yield (kilogram) and income ($) returns per unit of water, jobs per unit of water should also receive priority in crop choices.

3.4 Relooking water productivity measurements in crop production in the light of total factor productivity

Agricultural productivity is a measure of the performance of crops and other farm outputs such as trees, livestock, fish, wool, and poultry in terms of outputs and income. An increase in productivity indicates that inputs are being used more efficiently—that is, fewer inputs are required to produce the same output or, alternatively, that additional output is possible from a given level of input use. The inputs that can be considered are land, labor, water, and fertilizer, and pesticides (Sheng, Mullen, & Zhao, 2011).

Two types of measures of productivity are typically calculated. The first one is total factor productivity (TFP), which is a ratio of a measure of total output to a measure of multiple inputs used in the production process. The second is partial factor productivity (PFP), which is a ratio of a measure of total output to a measure of a single input category (Sheng et al., 2011).

There is insufficient clarity about the scientific basis for measuring water productivity in crop production and farming which implicitly attributes the productivity gains to efficient use of water and vice versa. The reason is that any changes in yield or income from crops can be a result of improvement in soil nutrient regime, choice of right crop technology, changes in labor inputs for farm management, besides irrigation water application and its efficiency. Hence, theoretically what is being measured is the partial factor productivity of water, for biomass output or income from individual crops or the entire farm.

Compared with TFP measures, PFP measures (such as labor productivity of crop, water productivity of crop) are of limited use for summarizing the overall productivity performance of the sector (Sheng et al., 2011). This is partly because PFP measures can result in a misleading assessment of farm productivity performance if the effects of change in inputs (fertilizer and pesticides), technology changes like IPM, and technological improvements embodied in other inputs (like in the case of seed technology) are (incorrectly) attributed to changes in only one particular input like labor or water.

In purely irrigated crops, in which the contribution of rainfall to biomass yield is almost absent (Fig. 3.1), the measurement of partial factor productivity of inputs such as fertilizers, labor, crop technologies, etc. would be of limited use, as this would lead to wrongly attributing all changes occurring in the crop yield and income as a result of changes in irrigation water input, to the use of other inputs mentioned. Let us assume that the partial productivity of fertilizer in generating income from the crop is γ. Such measurements would lead to a wrong inference that the yield can be enhanced by a factor of γ, by increasing the level of inputs by one unit, or the productivity in relation to these inputs can be enhanced merely by improving the efficiency of fertilizer use. The fact is that the effect of fertilizer and other inputs would be seen only if the crop receives the minimum amount of water for transpiration to reach maturity and produce yields. Here irrigation is the most critical input. But this effect would be limited. Any further increase in some of the inputs like nutrients to get higher yield will have to be compensated with the additional dosage of irrigation water for proper intake of these nutrients by the plants. Hence, partial factor productivity of water (which is also known as "water productivity") would make logical sense in such situations.

Conversely, it would be fallacious to measure the productivity of crop or farm in relation to irrigation water when the crop can be grown under rainfed conditions in the same locality with significant contribution of rainfall crop yield (Fig. 3.2), and attribute the performance of the crop in terms

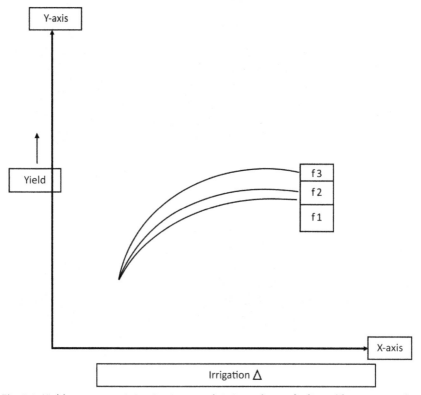

Fig. 3.1 Yield response to irrigation in a purely irrigated crop. f1, f2, and f3 correspond to yield obtained with three different levels of fertilizer dosage. *Source: Kumar, M. D. (2014). Changes in water productivity in agriculture with particular reference to South Asia: Past and the future. Final report submitted to International Water Management Institute, New Delhi.*

of yield and income entirely to irrigation water. In such situations, the effect of fertilizers, labor, crop variety, etc., would be significant determinants of productivity changes. Here, ideally, the partial productivity of various crop inputs (such as fertilizer, pesticide, labor, etc.) can be measured, depending on which factor of production is scarce.

Here, measuring the marginal productivity of irrigation water would be scientifically correct and practically useful, if the contribution of rain, fertilizer, and labor to crop yield and income changes can be segregated. This can be done if data on production and inputs are available for a sufficiently large sample of farmers, who represent a wide spectrum in terms of irrigation dosage—from no irrigation to full irrigation to meet the entire crop water requirement, all with more or less the same date of sowing, same variety, and

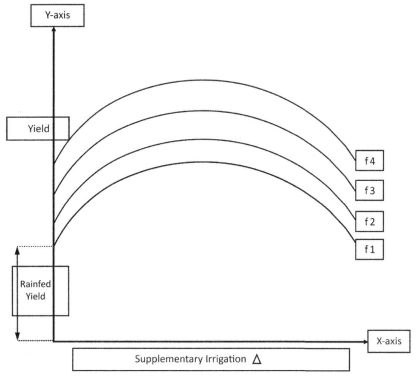

Fig. 3.2 Yield response to purely rainfed crop. f1, f2, f3, and f4 correspond to yield obtained with three different levels of fertilizer dosage. *Source: Kumar, M. D. (2014). Changes in water productivity in agriculture with particular reference to South Asia: Past and the future. Final report submitted to International Water Management Institute, New Delhi.*

with the same dosage of nutrient inputs to control all factors of production other than water input. Using this data, a single factor production function can be estimated with X values representing irrigation and Y values representing yield or income. The marginal productivity of irrigation water at different dosages of irrigation can be estimated along with average productivity, for different levels of irrigation, like X_1, X_2, X_3, and so on by estimating the slope of the production curve between these levels.

3.5 Production function, marginal and average water productivity, and technical and allocative efficiency

As the resources become scarce, producers seek ways to enhance the productivity of the resources and, of the entire production system (Gichuki et al., 2006). In the context of irrigated agriculture, when water becomes

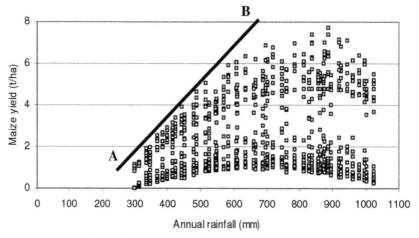

Fig. 3.3 Impact of rainfall on maize yield in Ewaso Ngiro Basin in Kenya.

scarce, understanding the crop production function becomes a precondition for identifying opportunities for improving the performance of a production system and enhancing the productivity of water.

Productivity increases can be achieved by two approaches: (a) by increasing technical efficiency through more efficient utilization of production inputs; and (b) increasing allocative efficiency by producing outputs with the highest returns. Gichuki et al. (2006) illustrated the ways to enhance the production efficiency, by using the yield function of maize in relation to rainfall.

The line A–B in Fig. 3.3 (source: NRM3, 2000) defines the limits of technical efficiency, which is the maximum level of outputs possible when the input resources are used most efficiently. Points below the curve are technically inefficient because the same level of yield could be attained with less amount of water. The different levels of yield for the same amount of irrigation water could be because of any of the following: different levels of agronomic inputs; changes in irrigation technology; changes in soil characteristics and quality; varietal differences or all of them.

To optimize the production system and to aid farmers in deciding the optimal level of water input to secure high outputs and water productivity, the concepts of "average productivity" and "marginal productivity of water" could be useful (Booker & Trees, 2020; Gichuki et al., 2006; Kumar, 2009). In order to illustrate this concept, a single factor production function, wherein the yield can vary only with changes in irrigation dosage, is presented in Fig. 3.4. It shows the range in crop production for a given level of technical efficiency and other inputs, which are kept constant.

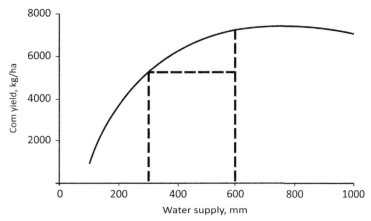

Fig. 3.4 Single factor production function. *Source: Frizzone, J. A., Coelho, R. D., Dourado-Neto, D., & Soliana, R. (1997). Linear programming model to optimize the water resource use in irrigation projects: An application to the Senator Nilo Coelho Project. Sci. Agric. (Piracicaba, Braz.) 54: 136–148.*

From Fig. 3.4, it is evident that the incremental output per unit input (marginal productivity of water, $\emptyset^1$) decreases at higher levels of input and later on becomes negative (Booker & Trees, 2020). A decrease in crop yield at higher levels of water input is mainly attributed to inhibited uptake of oxygen by the roots under water-saturated soil conditions (Gichuki et al., 2006; Kumar, Patel, Ravindranath, & Singh, 2008), and the excessive leaching of nutrients in the soils (Quezada, Fischer, Campos, & Ardiles, 2011).

At low levels of water input, the marginal product of water ($\emptyset^1$) is very high and higher than average productivity ($\emptyset$). But the average water productivity keeps increasing with increasing irrigation dosage, till it becomes maximum and equal to marginal productivity. Marginal productivity keeps decreasing thereafter, flattens, and then becomes negative. The point, where average productivity becomes highest, corresponds with the level to which farmers would like to irrigate the crop if enhancing WP is the main consideration. Fig. 3.4, which shows the crop-water production function for irrigated corn in Brazil, indicates that at point A, a yield of 7000 kg/ha was achieved with the application of 600 mm water. But at a dosage of 300 mm, a yield of 5210 kg/ha was obtained. The technically efficient irrigation dosage is 300 mm, as it corresponds to the highest average water productivity of 1.83 kg/m^3, whereas that corresponds to 600 mm is only 1.6 kg/m^3.

But, if yield maximizing is the main consideration, then the farmer would like to irrigate the crop till marginal productivity of water becomes zero (i.e., a little over 600 mm). However, maximizing water productivity becomes a consideration only when access to water is restricted (or supply is rationed) or water is priced volumetrically. Or else, farmers would be interested in yield maximization so long as it doesn't affect the net returns from crop production.

Farmers allocate water to their farms, which consist of several crops, livestock, trees, etc., rather than to a field having a single crop (Kumar & van Dam, 2013; Loomis & Connor, 1996). The concept of allocative efficiency is used to aid farmers to take decisions on water management when he has to allocate the limited water to many crops within the same farm, which competes for water use. If water productivity is a main consideration for the farmer, his preference would be to allocate all the water to the crop which gives the highest water productivity in $/m^3 of water, after exploring all the technical efficiency improvement measures which maximize the water productivity for the crop.

But the farmer also has to maximize the income. In that case, he/she also has to make sure that the maximum amount of land is put under the production of this particular crop to consume all the water. However, this may not always be possible, due to land constraints, and some water can still be left unused. Such a situation arises usually when the crop which gives the highest water productivity in monetary terms ($/m^3) produces very little net income per unit of land ($/ha), and the high water-productivity is mainly due to low water consumption (i.e., irrigation requirement, Δ), which reduces the value of the denominator. In that case, the farmer has to allocate some water to another crop, which gives a higher income return per unit of land, but is less water productive and requires more irrigation per unit of land (Δ).

This becomes an optimization problem. If we assume that the water productivity of crop A is $ 2.5/m^3 of water, and that of crop B is $ 1.0/m^3 of water, and crop A requires 2000 m^3 of water per ha producing a net income of US $ 5000, and crop B requires 8000 m^3 of water, producing a net income of US $ 8000. The total land area available for production is assumed to be 3 ha. Here, in this case, if the farmer allocates 2 ha for crop B and 1 ha for crop A, he/she would earn a profit of US $ 21,000, consuming 18,000 m^3 of water. At the same time, if he/she allocates 1 ha for crop B and 2 ha for crop A, he/she would gain an income of US $ 18,000, with total water use of 12,000 m^3. The water productivity is only US $ 1.16/m^3 in the first case, while it is US $ 1.5/m^3 in the second case. The farmer would normally prefer the second option, where he/she can use more water to gain more profit, though water productivity is low under that option.

3.6 Concluding remarks

There are several conceptual issues involved in the use of the terms "water use efficiency" and "water productivity" because of the definitions not being really holistic, and each one of them having the potential to convey different meanings to a different audience (plant physiologists, agronomists, farmers, irrigation engineers, economists/planners, policymakers, and water resources specialists) and in different contexts (like a field, farm, irrigation scheme, and river basin). While both the concepts have evolved over time and have different disciplinary origins and objectives, they are often used interchangeably creating confusion among the audience. Many definitions exist for each one of them, and new definitions have evolved in an attempt to capture complex situations and explain what parameters are used in the numerator and denominator of the functions used for estimating them.

While the concept of water productivity is more recent, there is rich international debate about different aspects of water productivity, starting from the usefulness of the concept of water productivity (Zoebl, 2006) to the determinants of water productivity to the scale of measurement of water productivity to the methods of measurement of water productivity to the water-saving impacts of water productivity improvement measures in agriculture. The discussion has become rich and complex with an attempt to integrate the concepts of total factor productivity in agriculture, and the production function, marginal and average water productivity, and allocative efficiency. There is no doubt that the use of all these concepts helps water managers identify the new opportunities for as well as limits to improving water productivity in irrigated agriculture.

References

Abdullaev, I., & Molden, D. (2004). Spatial and temporal variability of water productivity in the Syr Darya Basin, central Asia. *Water Resources Research*, *40*(8).

Aggarwal, P. K., Hebbar, K. B., Venugopalan, M. V., Rani, S., Bala, A., Biswal, A., & Wani, S. P. (2008). *Quantification of yield gaps in rain-fed rice, wheat, cotton and mustard in India.* (No. 43).

Ahmad, M. U. D., Masih, I., & Turral, H. (2004). Diagnostic analysis of spatial and temporal variations in crop water productivity: A field scale analysis of the rice-wheat cropping system of Punjab. *Journal of Applied Irrigation Science*, *39*(1), 43–63.

Barker, R., Dawe, D., & Inocencio, A. (2003). Economics of water productivity in managing water for agriculture. In J. Kijne, et al. (Eds.), *Water productivity in agriculture: Limits and opportunities for improvement. Comprehensive assessment of water management in agriculture* (pp. 19–35). UK: CABI Publishing in Association with International Water Management Institute.

Booker, J. F., & Trees, W. S. (2020). Implications of water scarcity for water productivity and farm labor. *Water*, *12*(1), 308.

Food and Agriculture Organization of the United Nations. (2003). *Unlocking the water potential of agriculture*. Rome: FAO.

Gichuki, F., Cook, S., & Turral, H. (2006). Agricultural water productivity: Issues, concepts and approaches. In *Basin focal project working paper 1 (no. 618-2016-41183)*.

Gosain, A. K., Rao, S., Srinivasan, R., & Reddy, N. G. (2005). Return-flow assessment for irrigation command in the Palleru River basin using SWAT model. *Hydrological Processes: An International Journal, 19*(3), 673–682.

Keller, A., Keller, J., & Seckler, D. (1996). *Integrated water resource systems: Theory and policy implications, IWMI Research Reports # 3*. Colombo, Sri Lanka: International Water Management Institute.

Kijne, J., Randolph, B., & David, M. (2003). Improving water productivity in agriculture: Editors' overview. In J. W. Kijne, R. Barker, & D. J. Molden (Eds.), *Vol. 1. Water productivity in agriculture: Limits and opportunities for improvement* (Cabi).

Kumar, M. D. (2005). Impact of electricity prices and volumetric water allocation on energy and groundwater demand management: Analysis from Western India. *Energy Policy, 33*(1), 39–51.

Kumar, M. D. (2009). Opportunities and constraints in improving water productivity in India. In M. D. Kumar, & U. Amarasinghe (Eds.), *Strategic Analyses of the National River Linking Project (NRLP) of India Series 5. Water productivity improvements in Indian agriculture: Potentials, constraints and prospects* (p. 163).

Kumar, M. D., & Patel, P. J. (1995). Depleting buffer and farmers response: Study of villages in Kheralu, Mehsana, Gujarat. In M. Moench (Ed.), *Electricity prices: A tool for groundwater management in India? Monograph*. VIKSAT-Natural Heritage Institute: Ahmedabad.

Kumar, M. D., Patel, A., Ravindranath, R., & Singh, O. P. (2008). Chasing a mirage: Water harvesting and artificial recharge in naturally water-scarce regions. *Economic and Political Weekly*, 61–71.

Kumar, M. D., Scott, C. A., & Singh, O. P. (2011). Inducing the shift from flat-rate or free agricultural power to metered supply: Implications for groundwater depletion and power sector viability in India. *Journal of Hydrology, 409*(1–2), 382–394.

Kumar, M. D., Singh, O. P., & Sivamohan, M. V. K. (2010). Have diesel price hikes actually led to farmer distress in India? *Water International, 35*(3), 270–284.

Kumar, M. D., & van Dam, J. C. (2013). Drivers of change in agricultural water productivity and its improvement at basin scale in developing economies. *Water International, 38*(3), 312–325.

Loomis, R. S., & Connor, D. J. (1996). *Productivity and management in agricultural systems*. Cambridge, NY: Cambridge University Press.

Meinzen-Dick, R. (1995). Timeliness of irrigation. *Irrigation and Drainage Systems, 9*(4), 371–387.

Molden, D., Murray-Rust, H., Sakthivadivel, R., & Makin, I. (2003). A water-productivity framework for understanding and action. In J. W. Kijne, R. Barker, & D. J. Molden (Eds.), *Vol. 1. Water productivity in agriculture: Limits and opportunities for improvement* CABI Publishing.

NRM3. (2000). *Natural resources monitoring, modelling and management database, Laikipia research program, Nanyuki, Kenya*.

Quezada, C., Fischer, S., Campos, J., & Ardiles, D. (2011). Water requirements and water use efficiency of carrot under drip irrigation in a haploxerand soil. *Journal of Soil Science and Plant Nutrition, 11*(1), 16–28.

Renaults, D., & Wallender, W. W. (2000). Nutritional water productivity and diets. *Agricultural Water Management, 45*(3), 275–296.

Sheng, Y., Mullen, J. D., & Zhao, S. (2011). A turning point in agricultural productivity: Consideration of the causes. In *ABARES research report 11.4, 4 for the Grains Research and Research and Development Corporation, Canberra, May*.

Singh, R., van Dam, J. C., & Feddes, R. A. (2006). Water productivity analysis of irrigated crops in Sirsa district, India. *Agricultural Water Management, 82*(3), 253–278.

Van Dam, J. C., Singh, R., Bessembinder, J. J. E., Leffelaar, P. A., Bastiaanssen, W. G. M., Jhorar, R. K., … Droogers, P. (2006). Assessing options to increase water productivity in irrigated river basins using remote sensing and modelling tools. *Water Resources Development, 22*(1), 115–133.

Zoebl, D. (2006). Is water productivity a useful concept in agricultural water management? *Agricultural Water Management, 84*(3), 265–273.

Zwart, S. J., & Bastiaanssen, W. G. (2004). Review of measured crop water productivity values for irrigated wheat, rice, cotton and maize. *Agricultural Water Management, 69*(2), 115–133.

CHAPTER 4

Estimating different productivity functions: Theory and review of past global attempts

4.1 Introduction

Depending on the scale and the stakeholder interests, different water productivity functions are used to assess performance irrigated areas (Gichuki, Cook, & Turral, 2006; Molden et al., 2007). As regards the scale of assessment, water productivity can be estimated at the plant level, at the plot level and field level, it can be measured at the farm level, at the system (scheme) level, and also at the watershed or basin level (Molden et al., 2007). Depending on the scale, the parameter to be considered in both the numerator (output or production-related term) and denominator (input or water-related term) for estimating water productivity would change. For instance, plant-level measurement of water productivity will have to consider the biomass yield or grain yield per unit of transpiration (T). Field level measurement of water productivity would essentially deal with biomass output or income per unit of irrigation water applied or depleted. At the farm level, since there will be several crops, allocative efficiency is important, and therefore to compare the performance, we will have to consider the net income per unit of water allocated to the farm.

Depending on the stakeholder interests, the determinants of water productivity would also change. For a plant physiologist, the amount of dry matter per unit of transpiration is what matters as the aim is to utilize the light and available moisture to maximize the biomass output and grain yield. For an agronomist, it is the yield per unit of evapotranspiration, with the aim to get sufficient food or biomass. For a farmer, it is the net income per unit of irrigation water applied that matters as the concern is to maximize the income and not the crop yield. For the irrigation engineer, it is generally

65

the income from the unit of water supplied from the scheme that matters as the concern is to have optimum water allocation. For the policymaker, the interest is in maximizing the economic output per unit of water available in the basin.

Estimating water productivity involves several challenges because of the complex nature of the output (production-related) and input (water-related) variables often involved (Molden et al., 2007). Depending on the scale, the indicators used for measuring the change in water productivity would also change along with changes in the degree of complexity involved in measurement. For instance, the indicator of improvement in agricultural water productivity at the river basin level can be simple and easily measurable, in the form of an increase in river flows or the amount of water remaining in reservoirs of the basin at the end of the hydrological year. At the same time, the measurement of changes in water productivity at the plant level (also known as transpiration ratio) can be quite difficult as there are no direct methods, and the indirect method involves estimation of leaf area index of the plant and soil evaporation, and measurement of evapotranspiration in the field (Kumar, 2014).

The chapter discusses the methodologies for measurement/estimation of various water productivity functions at the plant, plot/field, system and basin levels, synthesized from various published works of various scholars in the next section. A detailed discussion on the indicators for measuring the performance of crops and agriculture with regard to changes/improvement in water productivity at various scales is provided in the third section. The fourth section provides a review of select studies that deal with the measurement/assessment of water productivity in agriculture at various scales, with a subsection on the impact of mulching and climate change on crop water productivity. The last section offers certain concluding remarks.

4.2 Measurement of water productivity

Measurement of water productivity is scale-dependent (Dong et al., 2001; Loeve et al., 2004; Palanisami, Senthilvel, Ranganathan, & Ramesh, 2006). The determinants of water productivity (i.e., the parameters to be considered as numerator and denominator of the equation) and the unit of analysis of water productivity are determined by the purpose. It also depends on the concern of the stakeholder, who is interested in the performance assessment (Kumar and van Dam, 2009, 2013). Hence, depending on the objectives, the stakeholder involvement, and the concerns, the criteria for measurement of water productivity will change.

4.2.1 Plant level water productivity

For plant breeders, water productivity measurement at the plant level makes sense as the objective is to maximize the biomass output per unit of transpiration (Kg/T). But estimating transpiration values for crops is not easy and straightforward. It involves direct measurement of Leaf Area Index and then obtaining values of other parameters from established functions showing a relationship between bare soil evaporation, and evapotranspiration, and leaf area index.

Siddique, Belford, et al. (1989) and Siddique, Kirby, and Perry (1989) fitted curves using multiple regression analysis of measured values of leaf area index (LAI) and cumulative values of bare soil evaporation, E_s and evapotranspiration E_t to obtain weekly values for these parameters. Soil evaporation under each cultivar (E_{SC}) was then estimated using the coefficient for the total solar energy flux (Incident solar energy-light interception) from the values of bare soil evaporation as:

$$E_{SC} = E_S \, \mathrm{X} \, \alpha \qquad (4.1)$$

$$\alpha = e^{Kgr \, X \, LAI} \qquad (4.2)$$

where K_{gr} is the light extinction coefficient and LAI is the leaf area index.

Once, the values of E_{SC} are estimated for different time intervals, the corresponding values of transpiration can be worked out by subtracting these values from the corresponding E_t values (Siddique, Belford, et al., 1989; Siddique, et al., 1989).

4.2.2 Plot/field level water productivity

Agricultural scientists working on crop-water relationships would be concerned with water productivity at the plot or field level, and therefore the measure of water productivity would be kg of biomass produced per unit of water supplied to the plot or field. However, the determinant for measurement could be any of the following: the volume of water applied; total volume of water applied in the system, which includes the moisture in the soil profile; the total amount of water consumed by the crop in the field, i.e., evapotranspiration (ET); and finally, the total consumptive use (CU), which includes the nonbeneficial evaporation and nonrecoverable deep percolation (Allen, Willardson, & Frederiksen, 1997). The value to be used in the denominator would be decided by the objective of the assessment.

For irrigation engineers, whose objective is to examine how efficiently irrigation water is delivered in the field for raising crops, then the criteria for performance assessment would be the productivity of "applied water."

Here, the underlying premise is that for every unit of water applied, the crop consumptive use (ET) is highest, and the beneficial "return flows" as well as the nonbeneficial (evaporation and nonrecoverable deep percolation) components are reduced to a minimum.

This can be done by taking the ratio of the total crop output (Y) and the volume of water applied in the field (Δ) or the net return from crop production and the volume of water applied in the field. Obviously, if water is efficiently applied in the field, then the yield realization would be good, while the total dosage of water application would remain optimal, producing high productivity of applied water (∂) in kg/m^3 of water. But as discussed earlier, if higher efficiency in water delivery is achieved through an improved irrigation system with extra investment, then the net income per unit volume of water applied, i.e., combined physical and economic productivity of applied water ($\emptyset$) in $/m^3, should also be measured.

Plot level physical productivity of applied water as $\partial = (Y/\Delta)$.

Plot level combined physical and economic productivity of applied water as $(\emptyset) = (NR/\Delta)$.

Here, Δ is the volume of water applied per ha at the field inlet; and Y the total crop output in kg/ha; NR is the net return in $/ha.

As regards the measurement, the volume of water applied to the field can be measured using instruments such as partial flumes, V-notches (for channel flow), and current (water) meters for well irrigation.

If the objective of the measurement is to see how efficiently, the soil water along with irrigation water is used, then the productivity of total water applied in the system should be the consideration, and the same can be measured as:

Plot level physical productivity of total water in the system as $\partial = (Y/(\Delta + P_e))$.

Plot level combined physical and economic productivity of applied water as $(\emptyset) = (NR/(\Delta + P_e))$.

Here, P_e is the effective rainfall, or the water in the soil profile from rainfall, before the sowing operation.

If the total amount of water available in the soil profile from rainfall prior to sowing has to be measured, this would be possible through the use of soil tensiometers of a neutron probe. This can be added to the irrigation to arrive at the total water applied.

If the objective of performance assessment is to see how efficiently the water depleted (or consumed, CU) is utilized by the production system for realizing the yield or income, then water productivity should be measured in relation to the total amount of CU. Here, the underlying premise is that if water is efficiently used, for every unit of water depleted or consumed, the ET component will be highest, and nonbeneficial (evaporation and nonrecoverable deep percolation) will be minimal.

Such concerns of reducing CU without reducing the ET would be relevant in regions with high aridity and deep water-table conditions, where water could be lost in the vadose zone, without being recovered for reuse, or water from the soil exposed to sunlight (not covered by the canopy) between distantly spaced plants is lost in evaporation, due to inefficient method of irrigation (Kumar & van Dam, 2013).

The best way to estimate the "depleted water" (consumed water) in irrigated production is to use a soil water balance model, such as soil water atmosphere plant (SWAP) model. The model can to a great extent capture the soil evaporation and the soil moisture changes in the unsaturated zone along with root water uptake by plants, to estimate the total water depleted and the return flows. However, the SWAP model is data intensive and requires a wide range of data on soil hydraulic properties, plant root system characteristics, and agrometeorology to estimate the moisture changes in the crop root zone and unsaturated zone (moisture flux), evapotranspiration (which takes into account soil evaporation), and groundwater recharge. The model also estimates evaporation and transpiration separately (Feddes et al., 2001).

4.2.3 System level water productivity

The managers of irrigation systems would be preoccupied with the performance of the system in terms of the command area served, and therefore would be interested in knowing how beneficially the water stored in the irrigation reservoir or supplied from a running water source is used in the command. Here, there are many areas where water could be consumed for nonbeneficial uses. They are water lost in swamps; water lost in evaporation from the reservoir and canal system; water lost in deep vadose zones of aquifers, which is nonrecoverable by crops and trees; water lost in the field, mainly from evaporation after harvest.

Since the irrigation command generally would have many different crops and livestock, which the system feeds, the performance assessment would be based on economic criteria, i.e., Rs or $ per m³ of water. Here the denominator of the water productivity parameter should exclude all the water, which is available for reuse. This portion is known as the reusable fraction.

The reusable fraction would ideally consist of the water which recharges the groundwater, and the return flow into natural drains and which eventually gets captured in the downstream part of the basin in which the irrigation system operates. Such assessments would require complete water accounting for the irrigation scheme, which quantifies reservoir evaporation, evaporation loss from canals, evaporation from swamps along the estimation of total evapotranspiration from the crop-land irrigated by canal irrigation systems. But there are complex issues involved in this approach. Fields in many canal commands also receive water from wells, making conjunctive use systems. Because of this, it will be practically impossible to segregate the contribution of wells from the total output.

An alternative and a quicker method would be to look at the entire system, with gravity irrigation contributing to groundwater supplies and wells complimenting surface irrigation, and then quantify the total amount of return flows to groundwater and runoff into streams and drainage channels.

However, in the analysis of the performance of the irrigation system, the value generated from this water cannot be captured.

4.2.4 Basin level water productivity

Basin level water productivity measurements should take into account the actual amount of water depleted from the system in farming, manufacturing, and human uses and environmental management, against the total net value generated from various uses of water in the basin. Hence, basin-level water productivity shall be measured in $ per cubic meter of water.

Here, the estimation of the value of water used for human consumption and environmental management would require tools from environmental economics. While the production cost method could be employed in quantifying the value generated in human uses, methods such as "hedonic pricing" and "transaction cost method" could be used for estimating the value of water used for environmental management, comprising recreational uses, gardening, urban forestry, etc.

But estimating the amount of water depleted in various consumptive uses would require a proper "basin-wide water accounting" which takes into account the total water inflows from precipitation, the outflows for various competitive uses, and the change in storage. Ideally, the accounting has to be done for different hydrological years.

The outflows here would include the following: (1) consumptive use of water for biomass production (for crops and livestock); (2) evaporative loss of water stored in reservoirs for diversion to cropland and rangeland for agricultural production and fisheries; (3) evaporative loss of water in canals; (4) evaporation from the harvested fields and swaps; (5) consumptive water use in domestic and industrial sectors, which is exclusive of the return flows; and outflows into natural sinks (ocean, saline aquifers, etc.). The change in storage would essentially include (+ive) change in groundwater storage; and annual changes in reservoir storages. The continuity equation for inflows and outflows would help quantify the actual amount of water depleted in agricultural production. The inflows minus the positive changes in storage and the amount of water consumed in other sectors of the water economy would provide the quantum of water consumed in agriculture.

From the equation, it is evident that one way to enhance basin level water productivity is to capture the excess water which flows into the natural sink and divert it for beneficial uses such as crop production, manufacturing, environmental management services, etc.

4.3 Consumed fraction vs nonconsumed fraction

Chris Perry is one of the few water resource scientists globally who strongly argue for the need to make a clear distinction between water applied in irrigation and consumptive use, to capture scale effects in water productivity measurements, which is crucial while analyzing the performance of irrigation systems (see Perry, 2007). As pointed out by Perry (2007), the water applied (including irrigation water and available soil moisture) has two important components, viz., *consumed fraction*, which consists of beneficial consumptive use (ET) (a); and, nonbeneficial consumptive use (b)— evaporation from the capillary rise and ET from weeds; and, *nonconsumed fraction*, which consists of recoverable flows (c) (recharge from return flows to shallow aquifer which is pumped back, and return flows to drainage and streams, which is picked up downstream); and nonrecoverable flows, which consists of return flows to saline aquifers and discharge into streams which go uncaptured. While the hydrological components "a" and "c" are beneficial, in his opinion, the real water saving in irrigated crop production can come from a reduction in nonbeneficial consumed fraction (b); and nonrecoverable nonconsumed fraction (d).[a]

---

[a] Allen et al. (1997) treated the non-recoverable portion of the return flow as depleted water and therefore considered it as part of the consumptive use. In reality, this does not make any difference in assessing the performance of irrigation systems as this portion is treated as nonbeneficial. Kumar, Turral, Sharma, Amarasinghe, and Singh (2008) and Kumar and van Dam (2013), however, argue that deep percolation into vadose zone of aquifer should be treated as 'non-recoverable', when it is very thick. Hence, this will also have to be treated as consumptive use or depletion.

According to him, the lack of ability to segregate recoverable, nonconsumptive use from "nonrecoverable, nonconsumptive use" and nonbeneficial consumptive use has resulted in underestimating efficiencies in traditional irrigation systems, particularly large public irrigation systems, and overestimating the savings through "water-saving that is possible through efficient irrigation technologies. He cites four examples of pervasive outcomes from such a flawed approach.

They included (1) a water-short country in the Middle East investing in on-farm measures to increase measured "efficiency" from 40%–50% to 60%–70%, so as to release water for further expansion of the irrigated area, with the measurements to date showing that the improved technology resulted in increased water consumption along with increased crop yield; (2) a city in the United States offering to pay for the lining of irrigation canals in a neighboring irrigation district, so as to use the "saved water" for city water supplies and industries, with detailed analysis of the situation showing that at the basin level, about 80%–90% of water was already consumed with minimal potential "savings"; and, (3) wet seeding of rice, which while significantly reducing the volume of water applied to the field, keeps a larger area of the field fully wetted for a longer time period than under the traditional nursery system (Perry, 2007).

4.4 Basin water accounting framework

Real water saving is defined as the process of reducing nonbeneficial depletion and making the water saved available for more productive use. In situations where water is scarce, reducing nonbeneficial depletion becomes one of the main ways of reducing water scarcity. Reduction in nonbeneficial depletion can be accomplished through reducing: flows to sinks (Perry, 2007; Gichuki et al., 2006) and "nonrecoverable" deep percolation (Allen et al., 1997; Kumar & van Dam, 2013); and nonbeneficial evaporation (Gichuki et al., 2006; Kumar & van Dam, 2013; Perry, 2007). Hence, it is important to address the following key questions: What happens to the water that is lost through runoff and deep percolation, at a given study domain in a basin? What effect does reduction in nonbeneficial depletion have on systems that were dependent on the water it provided? What happens when the water is saved through reduced runoff and deep percolation?

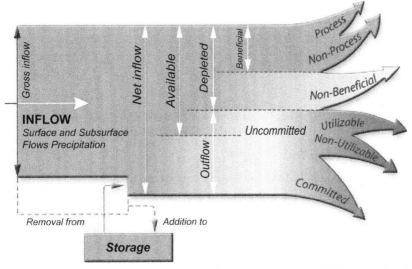

Fig. 4.1 Basin water accounting framework. *Source: Molden, D. J. (1997) Accounting for Water Use and Productivity, SWIM (System-wide Initiative in Water Management) Paper 1, Colombo: International Water Management Institute.*

The basin water accounting framework developed by Molden, Murray-Rust, Sakthivadivel, and Makin (2003) is useful in addressing such questions in any given domain within a basin (Fig. 4.1). It defines the following terms:

- *Gross inflow* is the total amount of water flowing into the study area from precipitation, rivers, and subsurface sources (groundwater).
- Net inflow is the gross inflow less any increases in storage in the surface soil or groundwater.
- *Available water* represents the amount of water available for use that is the net inflow minus the committed and nonutilizable outflow.
- *Water depletion* is the use of water within the system that renders it unavailable for further use. This has two distinct components, viz., beneficial depletion; and nonbeneficial depletion. Beneficial depletion can again be classified into process depletion, which is the amount of water diverted for use that is depleted to produce a human–intended product; and nonprocess depletion, which occurs when water is depleted, but not by a human-intended process. Nonbeneficial depletion occurs when water is depleted through evapotranspiration that is not beneficial.
- *Outflows* have two components, viz., committed outflow, which is that part of outflow from the study area that is committed to other uses such

as downstream environmental requirements or downstream water rights; and uncommitted outflow, which is that part of outflow that is available for use within the study domain, but flows out of the basin due to lack of storage or sufficient operational measures. Some of the uncommitted outflows can be nonutilizable.

4.5 Monitoring mechanisms and indicators to appraise improvement in water productivity at different scales

4.5.1 Plant level monitoring

When we look at an individual plant of a particular crop, the only beneficial use of water for biomass production is "transpiration." The relationship between transpiration and biomass output is linear, though can; however, be influenced by the nutrient and salinity levels in soils. But at the plot level, the ET becomes the beneficial use. The added component in the case of the latter is because evaporation from the land under the canopy is inevitable. Hence, the indicators used for measurement of water productivity should be the ratio of the yield and the total crop transpiration (Kg/Ta), or the inverse of it called the "transpiration coefficient" (Ta/kg). This essentially means that the only way to improve water productivity at plant scale is to manipulate biotic and abiotic factors which determine the yield, and the plant physiological process, which determines "transpiration."

There are major biotic and abiotic factors affecting crop growth and development, viz., radiation, temperature (yield-determining); pests and diseases (yield-reducing) (Aggarwal, Kalra, Bandyopadhyay, & Selvarajan, 1995); and water and nutrient (yield-limiting) (Aggarwal et al., 1995; Schmidhalter and Oertli, 1991).

As regards transpiration, it is governed by environmental conditions (solar radiation, humidity, temperature, and wind speed), crop duration, and plant physiology. Reduction in transpiration can be achieved by growing the plant when incident solar radiation is lower, temperature and wind speed are less, and humidity higher.[b] The second and third factors, i.e., crop duration and plant physiology, can be manipulated through plant genetics, which involves the selection of germplasm of plant varieties that mature within a shorter time duration or plant varieties that have a lower leaf area index.[c] But it is also important to reckon with the fact that manipulations

[b] Such as advanced or delayed sowing to increase the humidity or reduce the aridity, and reduce the temperature.

[c] Leaf area index is a measure of the cumulative leaf area from top of canopy to bottom and the ground area.

such as choosing short duration varieties or raising plants when solar radiation is lower, might have negative implications for yield.

Hence, a simple indicator for assessing the potential change in transpiration is the duration of plant growth. A shorter duration indicates lower transpiration if the environmental conditions under which the crop is raised do not change. However, if the latter happens, to incorporate its influence on plant transpiration, the actual crop transpiration itself needs to be estimated, by studying the plant architecture (leaf area index, stem height, etc.) and agrometeorological parameters. For instance, in many parts of western, northwestern, and central India, pearl millet, and sorghum are grown during summer as well as monsoon. The transpirative demand of the crop during summer would be much higher than that of monsoon, due to differences in incident solar radiation, which will also affect the yield potential. Hence, the second indicator is the environmental conditions during the growing period, expressed as daily transpirative demand. The third is the leaf area index.

The indicators of improvements in the potential yield of plants could be the following: the successful development of crop varieties in regions that receive higher solar radiation; development of drought and disease-resistant varieties; and, development of varieties that have higher efficiency of use of nutrients and water from the soil profile, by virtue of the rooting system characteristics of the plants.

Mechanisms for monitoring plant-level water productivity are field instrumentation for measuring the environmental factors such as solar radiation, temperature, wind speed, and humidity; duration in days under different growing stages of the plant; physical measurement of leaf area index of the plant and plant height which would help estimate the actual crop transpiration using weather data; and measurement of actual crop yield at the plot level. For grain yield, both leafy biomass (dry matter) and grain (fruit) yield need to be measured. Direct measurement of plant transpiration using a micro lysimeter can also be resorted to.

4.5.2 Field level monitoring

The indicator of field level applied water productivity improvements are provided by the changes in crop yield in relation to the total amount of water applied. Since crop yield is a function of transpiration, what determines the performance here is how much of the applied water is converted into beneficial transpiration.

However, water productivity has to be estimated in relation to ET, when the scope for recycling of water applied in the field is high, to see how the yield changes with ET. This can be done through direct measurement of ET using a lysimeter installed in the field. It can also be estimated through the estimation of crop ET using the solar energy balance method with the help of remote sensing-based tools, and direct measurement of yield from sample plots. Since ET can vary widely between fields depending on the variation in water stress experienced by the crops during the growth stage and hence the crop yield, scheme-wide mapping of ET and yield are essential to estimate the average values. But what determines the performance is what proportion of the evapo-transpiration is actually transpiration, and how much of it is soil evaporation.

The field-level water productivity (kg/ET) can be manipulated by improving the harvest index (Siddique et al., 1990); phenological stages of the plants; and the ratio of evaporation to transpiration. Raising the harvest index raises the grain yield per unit of ET. Faster development and earlier flowering allow a greater proportion of vegetative water use to occur at times of low vapor pressure deficit, for winter crops. Whereas, the number of leaves, their orientation, and leaf area index together deter-mine light interception and shading of the soil surface, which in turn, will influence the ratio of soil evaporation to evapotranspiration (ET) (Cooper, Keatinge, & Hughes, 1983; Tanner & Sinclair, 1983). Hence, while the ultimate indicators of improvements in field-level water productivity are yield, and ET/ water applied, in order to know the factors driving water productivity improvements, the following parameters also needs to be evaluated: harvest index; soil evaporation-evapotranspiration ratio (water productivity in relation to ET); and ET-irrigation ratio or con-sumptive fraction (for applied water productivity).

Siddique, Tennant, Perry, and Belford (1990), based on a study of Med-iterranean wheat varieties showed that substantial improvement of WUE, occurred in modern cultivars (46%) when compared with very old cultivars. This increase in WUE was associated with improvement in harvest index, and early flowering in modern cultivars which reduced the evaporation/ET ratio significantly (Perry and d'Antuono, 1989; Siddique, Belford, et al., 1989; Siddique, Kirby, and Perry, 1989). The study also shows that dry mat-ter WUE has changed very little, suggesting less improvement in biomass through breeding (Fig. 4.2).

4.5.3 System level monitoring

Measurement of system-level water productivity in relation to applied water is rather straightforward. It only has to take into account the amount of water utilized from reservoirs/diversion systems + the groundwater pumped for irrigation. Continuous monitoring of reservoir releases and groundwater

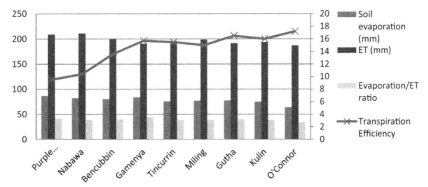

Fig. 4.2 Evaporation-ET ratio and transpiration efficiency in wheat. *Source: Siddique et al., 1990.*

pumping is, therefore, important. Reduction in release of water from reservoirs or pumped groundwater with no changes in the value of outputs, under conditions of prices remaining constant, would indicate an improvement in applied water productivity at the system level. This can happen through improvement in physical efficiency. This would not always mean that water is saved at the system level. Instead, it is quite likely that less amount of water returns to the shallow aquifer or the drainage channels/streams downstream. However, if such improvements in applied water productivity occur through improvement in allocative efficiency, with farmers shifting to low water consuming crops, this may lead to real water saving at the system level.

But in situations where the recycling potential of the water applied in the field is significant, then water productivity needs to be estimated in relation to the actual amount of water depleted. In that case, the measurement has to take into account the return flows from irrigated fields into shallow aquifers and the drainage canals and streams/rivers, along with the total amount of water utilized from the reservoir (including the evaporation) and the total amount of groundwater pumped for irrigation in the command area, to arrive at the value of the denominator. Ideally, water productivity estimated in relation to depleted water would be far higher than applied water productivity.

4.5.4 Basin level monitoring

Mechanisms for basin-level water productivity monitoring are hydrological monitoring of basin inflows (precipitation and stream-flows), groundwater level fluctuations and reservoir water level fluctuations; application of remote sensing tools for studying changes in land use and ET from cropland over time; and socioeconomic surveys to estimate annual agricultural (crop, livestock, and fisheries) and manufacturing outputs and water supply services being provided to rural areas and cities. Such long-term monitoring of basins and catchments would be quite a time-consuming and resource-intensive.

Indicators for basin-level water productivity changes are (1) outflows from the drainage outlet from stream gauges; (2) land-use changes (changes in the area under different crops, under rain-fed and irrigated conditions); (3) changes in consumptive use of water in urban and rural areas for domestic uses, livestock and manufacturing; and (4) changes in the proportion of the water allocated to high-value crops, and uses. The third component can be estimated by measuring the total water supplied to these sectors and the wastewater return flows.

A positive change in groundwater balance and enhanced reservoir storage in the basin over time without much changes in the inflows, and with no reduction in the value of the farm and manufacturing outputs that are dependent on water (under the condition of prices remaining constant) would indicate enhanced water productivity in the basin. Increased utilization of flows from the basin over time with a good proportion of the arable land remaining uncultivated due to lack of access to irrigation for production would indicate reducing basin-level water productivity. In that context, it should be kept in mind that in many regions of the world, there has either been a reduction in groundwater recharge due to drastic reduction in irrigated paddy area as noted by UNDESA (2016) in the case of Japan, or a rise in groundwater levels owing to the reduction in natural vegetation in the catchment, as noted by Leblanc, Tweed, Van Dijk, and Timbal (2012) in the case of Murray Dariling basin. Such changes have nothing to do with agricultural water productivity improvement. As noted by Masumoto, Hai, and Shimizu (2008), paddy fields commonly serve both agricultural and flood-control functions in low-lying areas in Japan, Cambodia, Vietnam, and other Asian countries.

4.6 Review of past research involving assessment of water productivity

Over the past 30 years, several studies were carried out by plant scientists, applied hydrologists, and agricultural economists on water use efficiency and water productivity in agriculture. These studies had different units of analysis—from plant level to plot level to system level to basin level.

The important findings from them are discussed in this section. A separate subsection offers a detailed review of the studies that analyzed the impact of mulching on crop water productivity, as available from a few Asian countries viz., India, Pakistan, and China.

Siddique et al. (1990) measured water use and water use efficiency of old and modern wheat cultivars and one barley cultivar in a Mediterranean environment. Water use efficiency for grain increased substantially from old to modern cultivars—from a lowest of 0.56 kg/m^3 for one of the old cultivators to the highest of 1.05 kg/m^3 for one of the modern cultivars. Water use efficiency for dry matter was similar between cultivars. Barley had the highest water use efficiency of grain (ratio of grain yield and actual ET) of both grain and dry matter (1.15 kg/m^3). Improved water use efficiency for grain in modern cultivars was associated with faster development, earlier flowering, improved canopy structure and higher harvest index. Modern cultivars used slightly less water than old cultivars.

Soil evaporation estimates showed that modern cultivars had lower rates of soil evaporation in the early part of the growing season. This was associated with their faster leaf area development and improved light interception. About 40% of the total water use was lost by soil evaporation with very little difference between wheat cultivars. Barley had 15% less soil evaporation than wheat. Water use efficiency for grain, based on transpiration (i.e., transpiration efficiency) increased from old cultivars to modern cultivars (from 0.95 kg/m^3 of water to 1.7 kg/m^3 of water. Transpiration efficiency for the four modern cultivars of wheat was 1.58 kg/m^3.

Molden et al. (2003) provided a basin water accounting framework to help understand the denominator used in water productivity at the field, farm, system, and basin levels. The framework recognizes depleted water as the one unavailable for further use in the hydrological system and included water evaporated, flows to sinks, and incorporation into products. For the treatment of water productivity at the system or basin level, it considers return flows from irrigated fields as fully "available for reuse" unless it is too polluted (see Molden et al., 2003: page 3). Seckler, Molden, and Sakthivadivel (2003) further expand the term "natural sink" to include two situations: (1) outflows of water from irrigated areas in deserts that subsequently get evaporated and (2) where severe mismatches between water supply and demand occur in terms of specific time and place (Seckler et al., 2003: page 3).

Ahmad, Bastiaanssen, and Feddes (2002) used the soil-water-atmosphere-plant (SWAP) model to estimate water flux in the unsaturated soil profile of groundwater irrigated areas of Pakistan Punjab under rice-wheat system and cotton-wheat system. It showed that the deep percolation (recharge) in irrigated fields cannot be estimated using root zone water balance as it will not be the same as the return flows from the plant root zone. The study quantified the moisture changes in the unsaturated soil profile during crop seasons, made the distinction between "process depletion" (transpiration) and evaporation from cropped land. The study found that the vertical water flux in the unsaturated zone is continuous under the rice-wheat system with frequent and intensive irrigation.

A study on water productivity of wheat in the canal irrigated areas of western Indo-Gangetic plains in Indian and Pakistan Punjab shows the improvement in water productivity due to both improvements in farm management practices—crop technology, timeliness of input use, and improvements in water management practices (Hussain, Sakthivadivel, Amarasinghe, Mudasser, & Molden, 2003). A similar study done by Bastiaanssen, Mobin-ud-Din, and Zubair (2003) in canal command areas of the Indus basin in Pakistan shows a positive correlation between yield and water productivity of wheat for both depleted water and diverted water wherein they considered evapotranspiration as the depleted water.

Singh, van Dam, and Feddes (2006) estimated salt and water balance at the farm level in Sirsa Irrigation System at Haryana. They used the SWAP model, based on Richards' equation for this. The soil hydraulic functions to be used as model parameters in SWAP were estimated, or in other words, the model calibration was done, through inverse modeling using pedotransfer functions, with measured values of soil moisture and the salt context in the soil for various time intervals. The model was later on validated using another set of measured values of soil moisture in the same fields for a subsequent set of time intervals. The soil water balance (change in soil moisture at a given depth at a given time) and water management response indicators, such as relative transpiration (T), rainfall and irrigation contribution to ET, percolation index, and salt storage index, for paddy-wheat and wheat-cotton systems, were estimated using the validated model.

An analysis of soil water balance in rice-wheat fields in Sirsa district of Haryana was carried out by Singh (2005) using the soil-water-atmosphere-plant (SWAP) model. It showed that the total water applied was in excess of the estimated ET (in the order of 290 mm to 561 mm). Interestingly, the ET value was higher for the field which had a lower dosage of irrigation. It shows that there is an ample opportunity for real water saving through a reduction in nonbeneficial E of ET and the part of soil moisture storage change, which would eventually get evaporated from the field.

By reducing irrigation dosage in such conditions as cited above, the farmers gain both higher land productivity (return per unit of land) and higher return per unit of water.

Zhu, Giordano, Cai, and Molden (2004) as a part of a water accounting exercise for the Yellow River basin, estimated water productivity (both physical productivity and economic efficiency) for many crops. Among all the three cereals compared, the physical productivity of water was highest for maize (1.40 kg/m^3) followed by wheat (0.59 kg/m^3), and lowest for rice. The economic efficiency of water was highest for cotton (\$ 0.19 m^3), followed by maize (\$ 0.15 m^3). They used the total volume of water delivered at the field inlet as the denominator in estimating water productivity functions.

Ahmad, Masih, and Turral (2004), estimated the spatial and temporal variations in water productivity (physical and economic) separately for process evaporation, soil evaporation, and actual ET which were estimated using the SWAP model for the rice-wheat area in Punjab. They found among others that the applied water (sum of precipitation and irrigation) far exceeded the evapotranspiration demand (ET) in the case of rice causing deep percolation, whereas it fell short of the ET requirements in the case of wheat, with some of the requirements being met by soil moisture depletion. They also found that the process depletion (transpiration) to produce a unit weight of cereal was slightly lower for rice when compared with wheat.

Molle and others (2004) in their study of water use hydrology and water rights in a village in Central Iran emphasize how the surface water flows (canals, river flows, etc.) and groundwater flows are inter-related when basin moves toward "closure," with storage, conservation, diversion, and depletion of water at one point determine what is available at another and therefore the interconnectedness of various users/actors through the hydrological cycle (Molle, Mamanpoush, & Miranzadeh, 2004). They argued that well development was tantamount to the reallocation of water from *qanat* owners to well owners. Also, the development of wells reduced groundwater flow for downstream users. Two aspects of the study are crucial from the point of view of analyzing system-level water productivity: (1) increasing efficiency of water from surface systems would have an adverse effect on groundwater availability when systems are hydraulically interconnected, and (2) pumping of water from the local aquifer can ensure higher reuse of water from canal thereby achieving optimal efficiency of use of surface water.

Kendy, Molden, Steenhuis, Liu, and Wang (2003) carried out a water balance approach to analyze the impact of policy interventions to affect sustainable water use in the semihumid north China plains. They used the difference between irrigation return flow [defined as precipitation + irrigation (I)–ET] and groundwater draft (I) as the net groundwater storage change. In their analysis, the entire return flow was treated as recharge to the aquifer system, which made them argue that any intervention to improve the physical efficiency of water use in crop production in the region, which eventually reduces return flow, would fail to make an impact on groundwater. Their analysis treated crop consumptive use (ET) as "water depleted" (Kendy et al., 2003) and did not consider the losses in the unsaturated zone, which can be significant when it is very thick (Kumar & van Dam, 2013). It is recognized that the ET values themselves could reduce with irrigation and soil management (Burt, Howes, & Mutziger, 2001), especially mulching that suppresses nonbeneficial evaporation from the soil (Wang et al., 2019; Wang, Jin, Šimůnek, & van Genuchten, 2014; Xie, Wang, & Li, 2005), therefore, improving groundwater balance. The significance of achieving better groundwater balance through a reduction in irrigation water application would increase with increasing inefficiency of conveyance of percolating water from the crop root zone to the groundwater system.

The past research on water productivity in agriculture had mainly dealt with the analysis of water productivity in certain field crops such as paddy, wheat, and maize and cash crops such as cotton at the field and regional level (see, for instance, Zwart & Bastiaanssen, 2004). They have not captured water productivity in composite farming systems and the key factors that cause water productivity improvements.

Kumar (2010) developed methodologies for assessing water productivity of irrigated crops (which also receive rainwater) and purely rainfed crops and estimated crop water productivity for several crops in several agroclimatic regions of the Narmada river basin, based on field data from several farms located in different agroclimatic zones. The analysis covered both physical and combined physical and economic productivity of water. For irrigated crops, which also received rainwater during the initial stages of crop growth, an analytical tool was developed to segregate the green water contribution of the yield from the total yield. The study also mapped the variation in water productivity of certain crops across agro climates. The study found that the productivity of blue water is far greater than the productivity of green water used in crop production in economic terms.

Kumar, Scott, and Singh (2011) assessed the crop water productivity for several crops in sample farming systems of South Bihar, eastern UP, and north Gujarat under well irrigation and also assessed the water productivity of the farming systems which consisted of crops and dairy farming. The analysis used productivity of applied water (biomass output and net income per unit volume of water applied to the crops and allocated for livestock), and compared overall farm-level water productivity (Rs/m^3 of water) under different cost regimes for irrigation water, by considering the following four categories (i) farmers who used electric pumps and do not pay for electricity for pumping groundwater; (ii) farmers who used electric pumps and pay for electricity on pro-rata basis; (iii) farmers who used diesel engines and pay for every unit of diesel used; and (iv) farmers who bought water from electric and diesel well owners on an hourly basis. For dairy production, the entire life cycle of the animals was considered for estimating the inputs and outputs. The water input for dairy production is considered the water embedded in animal feed and fodder plus the voluntary water consumption by animals. The study found that farmers who are confronted with a positive marginal cost of using irrigation water (electric well owners who paid on pro-rata basis, diesel well owners, and water buyers) obtained higher water productivity in economic terms than farmers who incurred zero marginal cost of pumping irrigation water.

4.6.1 Impact of mulching on crop water productivity

Among the water conservation measures, mulching has gained popularity because of its ability to reduce the rate of water loss (nonbeneficial consumptive use of water) from both surface and subsurface soils, while the ability of micro irrigation systems to save water in agriculture is somewhat controversial (see Frederiksen & Allen, 2011; Frederiksen, Allen, Burt, & Perry, 2012; Gleick, Christian-Smith, & Cooley, 2011; Kumar & van Dam, 2013). Mulch restricts the transport of water vapor from the soil surface to microclimate, which diminishes the direct evaporation loss of soil water (Xie, Wang, Jiang, & Wei, 2006; Yuan, Lei, Mao, Liu, & Wu, 2009) and increases the availability of soil water to the crops (Fuchs & Hadas, 2011). Straw mulching promotes transpiration from wheat crop fields by 14%–15% over bare conditions (Singh, Eberbach, Humphreys, & Kukal, 2011). In India, crop residues are mainly used as mulch material to conserve profile soil water (Chakraborty et al., 2008), but first-rate of decomposition reduces its impact at the later part of crop growth (Chakraborty et al., 2010; Fuchs & Hadas, 2011). The adoption of plastic mulch reduced the magnitude of evaporation loss by 55% over bare conditions; however, it enhanced the value of evapotranspiration (Xie et al., 2005). Organic and plastic mulches are widely used in vegetable production for various reasons including conservation of soil moisture (Kere, Nyanjage, Liu, & Nyalala, 2003).

Ngangom et al. (2020) conducted field experiments for two consecutive years at the North-Eastern Indian Himalayan region Umiam, Meghalaya to assess the effect of soil moisture conservation measures on soil and water productivity of different rainfed maize (*Zea mays* L.) – based cropping sequences. They observed a significant effect of mulching on WUE of all winter season crops. The highest water use efficiency (WUE) was recorded under in-situ maize stover mulch (MSM) + fresh biomass of white hoary pea (WHP) mulch followed by that under in-situ MSM + fresh biomass of ragweed (RW) mulch treatment in all crops. The water productivity (WP) was significantly influenced by mulch treatments in all cropping sequences. The highest pooled WP was recorded in the maize-rapeseed sequence followed by maize-black gram for all mulch treatments. Significantly higher WP was obtained with maize-rapeseed sequence under in-situ MSM + WHP mulch relative to no-mulch. The WP of the maize-rapeseed system under in-situ MSM + WHP mulch was 7% higher than that under no mulch. Similarly, WP obtained under in-situ MSM + WHP mulch and in-situ MSM + RW mulch was significantly more than that under no-mulch.

Singh, Singh, and BHATT (2014) conducted a field experiment with farmers' participation to investigate the performance of black plastic mulch on earliness, yield, water productivity, and production economics of summer squash, tomato, and capsicum under midhigh hills of district Chamoli (Uttarakhand) during the years 2009 and 2010. They found that the actual water applied was lower in mulch treatments and the water requirement was almost 30%–40% less than unmulched treatments of all three crops. The highest water saving was recorded in tomato (38.5% and 40.40%) followed by summer squash (32.6% and 31.5%) and capsicum (27.8% and 30.1%) during the first and second years, respectively. It was probably due to the reduced wet evaporating surface and greater foliage spread of all three crops which greatly reduced the moisture losses from the soil surface. From the data, it was evident that the use of black plastic mulch has enhanced the water use efficiency (WUE) and water productivity (WP) to a great extent for all the three crops under investigation in DP as compared to bare field cultivation. The black plastic treatment not only reduced the water requirement but also increased yield significantly. Thus, WUE was almost doubled or more in all three crops during both the years of investigation. Water productivity (WP) is also enhanced under mulch as less water is required to produce a unit weight of the crop output.

Zhang, Wei, Han, Ren, and Jia (2020) evaluated the effects of different film mulching methods on soil water, water productivity, crop yield, and economic benefits in a semiarid area of southern Ningxia, China. After 4 years of continuous treatment, the results indicated that maize planted in furrows separated by alternating large and small plastic film-mulched ridges (D) and maize planted in furrows separated by consecutive plastic film-mulched ridges (S) treatments significantly increased soil water content at depths of 0–200 cm during the early growth stage, but the increases in the soil water content were lower in the later stages of the crop growth process. On average, mulching increased grain yield by 31.3%, and biomass accumulation by 22.5%. Across four seasons, film mulching significantly ($P < 0.05$) increased water use efficiency and precipitation use efficiency, especially in D by 34.6% and 43.1%, and S by 35.3% and 42.8%, respectively. Mulching changed the crop water consumption characteristics and resulted in higher crop water productivity due to an improved hydrothermal balance, especially during the vegetative (30–90 days) and prereproductive (90–120 days) growth stages, which might explain the +ive yield effects in this system.

Zhang, Zhang, Qi, and Li (2018) evaluated the effect of plastic film mulching on improving crop yields on the Loess Plateau of China. They found that maize production increased significantly under plastic mulching (PM). The average grain yields across the Loess Plateau with plastic mulching were 58% higher than that in unmulched management (CK), at 5318.6 ± 1651 kg ha^{-1} in PM and 3370.1 ± 1680 kg ha^{-1} in CK. The higher production in PM compared with CK could be due to significantly greater water use efficiency (WUE) in PM than in CK. The simulated average WUE in the study area was 31 kg ha^{-1} mm^{-1} in PM and 20 kg ha^{-1} mm^{-1} in CK.

Jabran et al. (2015) conducted a study for two successive years (2008 and 2009) at the Agronomic Research Farm, University of Agriculture, Faisalabad, Pakistan to assess the potential role of mulches (plastic and straw) in improving the performance of water-saving rice production systems in comparison with no mulch used and conventionally irrigated transplanted rice. Water-saving rice production systems in this study comprise aerobic rice and transplanted rice with intermittent irrigation. These systems saved water (18%–27%) with improved water productivity more than the conventional system. They found both plastic and straw mulches helped improve moisture retention and water productivity (0.18–0.25 kg grain/m^3 water) relative to nonmulch treatments (0.19–0.29 kg grain/m^3 water). Mulch application was also helpful in reducing the number of nonproductive tillers and sterile spikelets and improving the productive tillers, kernel number and size, and kernel quality. They concluded that plastic mulch was more effective than straw mulch in improving water retention, water productivity and reducing spikelet sterility.

4.6.2 Climate change impact on crop water productivity

4.6.2.1 Impact on crop yield

It is well established that climate significantly affects crop yields. The yield of cereals has been reported to decrease under different future climatic scenarios (Parry et al., 1999). The Food and Agriculture Organization (FAO) and Intergovernmental Panel on Climate Change (IPCC) have estimated a reduction in cereal production for India by 125 m. ton with an overall increase in temperature by 2°C (Gahukar, 2009).

The likely impacts of climate change on crop yield can be determined either by experimental data or by crop growth simulation models (Kang, Khan, & Ma, 2009). A number of crop simulation models, such as CERES-Maize (Crop Environment Resource Synthesis), CERES-Wheat, SWAP (soil–water–atmosphere–plant), and InFoCrop, have been widely used by Kang et al. (2009) to evaluate the possible impacts of climate variability on crop production, especially to analyze crop yield-climate sensitivity under different climate scenarios. Kang et al. (2009) note that the studies about wheat production affected by climate change are mainly concerned with future CO_2 concentrations.

According to Hundal and Kaur (2007), an increase in the concentration of CO_2 up to 600 can mitigate the adverse effect of a rise in temperature by 1°C in rice, according to Sharma et al., (2013), an increase in concentration up to 682 ppm can mitigate the effect of temperature rise of 1.6°C in maize. Guo, Lin, Mo, and Yang (2010) reported inconsistent effects of climate change on wheat and maize yields in north China plains. Under the same scenario of atmospheric warming, wheat yield ascended, but maize yield descended.

Ortiz, Sayre, and Govaerts (2008) argued that wheat could adapt to the anticipated change in climate in the Indo-Gangetic Plains for the 2050s and suggested that global warming is beneficial for wheat crop production in some regions, but may reduce productivity in critical temperature areas. Anwar et al. [49] used CropSyst version-4 to predict climate change impacts on wheat yield in south-eastern Australia, and their results show that the elevated CO_2 level can reduce the median wheat yield by about 25%. Eitzinger, Stastna, Zalud, et al. (2003) used the CERES-wheat model to assess climate change impacts on wheat production under four climate scenarios, and the results show that the CO_2 effect will be mainly responsible for increasing crop yield. Luo, Williams, Bellotti, et al. (2003) discussed climate change impacts on wheat production with DSSAT 3.5 (Decision Support System for Agrotechnology Transfer) CERES-Wheat models under all CO_2 levels in Southern Australia for the 2080s, and the result shows that wheat yield would increase under all CO_2 levels, and the drier sites are more suitable for wheat production but are likely to have lower wheat quality.

Cuculeanu, Tuinea, and Balteanu (2002) predicted rainfed maize yield using CERES-Maize using CCCM and GISS climate models, which showed an increase in the dry matter by 1.4–2.1 ton/ha, and 3.5–5.6 ton/ha, respectively. Walker and Schulze (2006) used the CERES-Maize model to predict crop production in smallholders under different climate scenarios using the Mann-Kendall nonparametric test in South Africa, and the result showed that increasing inorganic nitrogen and rainwater harvesting can increase crop yield for smallholders in the long run.

Tojo Soler, Sentelhas, and Hoogenboom (2007) analyzed the impacts of planting dates and different weather on maize production in Brazil with CERES-Maize, and the result shows that a later planting date will decrease the average yield of maize by 55% under rainfed conditions and 21% under irrigated conditions. Popova and Kercheva (2005) analyzed the maize yield under precise irrigation and deficit irrigation for a 30-year period in Sofia Bulgaria with CERES-Maize, and the result showed that average productivity under deficit irrigation would be 60% lower than that under a sufficient soil moisture condition. Akpalu, Hassan, and Ringler (2008) studied the climate impacts on maize yield in the Limpopo basin (South Africa) and showed that increased temperature and rainfall had a positive effect on crop yield, and the rainfall had a greater effect on crop yield than temperature.

Droogers, van Dam, Hoogeveen, et al. (2004) studied climate change impacts on rice yield in seven basins with the SWAP and HadCM3 climate model under A2 and B2 scenarios, and the result shows that rice yields are expected to increase by around 45% and 30% for A2 and B2 scenarios, respectively. Krishnan, Swain, Bhaskar, et al. (2007) analyzed the impacts of elevated CO_2 and temperature on irrigated rice yield in eastern India by ORYZAI and InFoCrop-rice models, and the result shows that increased CO_2 concentration can increase the rice yield, and an increase in temperature can reduce it. The study suggests that the limitations on rice yield imposed by CO_2 concentrations and temperature can be mitigated, at least in part, by altering the sowing time and the selection of genotypes that possess higher fertility of spikelets at high temperatures.

Yao, Xu, Lin, et al. (2007) analyzed CO_2 level impacts on rice yield with the CERES-Rice model in Chinese main rice production areas, which shows that rice yield will increase with the CO_2 effect, otherwise it will decrease. Challinor and Wheeler (2008) used the GLAM (general large-area model) to analyze climate uncertainty impacts on peanut yield, and the result is that the yield can rise by 10%–30% with fixed-duration simulation.

Parry et al. (1999) used the IBSNAT-ICASA (International Benchmark Sites Network for Agrotechnology Transfer-International Consortium for Agricultural Systems Applications), dynamic crop model, to estimate climate potential changes in the major cereals and soybean crop yield, which showed that climate change would increase yields at high and mid-latitudes and decrease yields at lower latitudes.

Reddy and Pachepsky (2000) validated soybean yield prediction based on the GCMs and soybean crop simulator, GLYCIM in Mississippi Delta, providing a practical method to derive the general relationship between crop yields and climate change including temperature, precipitation, and CO_2 concentration. Challinor, Wheeler, Craufurd, et al. (2007) discussed the temperature effect on the crop yield in India with the regional climate model PRECIS and the GLAM crop model under past (1961–1990) and future (2071–2100) climate conditions. The result shows that the mean and high temperature are not the main factors that decide crop yield, but the extreme temperature will have a negative effect on crop yield when irrigation water is available for the extended growing period.

Popova and Kercheva (2005) note that the impacts of climate change on crop yield are different in different areas—with an increase in some regions and a decrease in some others. This difference in nature of impact is concerned with the latitude of the region and irrigation application. The crop yield can increase with irrigation and precipitation increases during the growing season; meanwhile, crop yield is more sensitive to precipitation than temperature.

Lal (2005) note that the positive effects of climate change on agriculture are concerned with the CO_2 concentration to augment, crop growth period increases in higher latitudes and montane ecosystems; the negative effects include the increasing incidence of pests and diseases, and soil degradation owing to temperature change.

Synthesizing these studies, it can be summarized that temperature increases will have a differential effect on the yield of cereals, with a change in latitudes—from positive for high and medium altitudes to negative in low altitudes. Further, the adverse effect of temperature on the crop can be offset by an increase in CO_2 in the atmosphere which is associated with global warming, as studies on wheat, paddy and maize have shown. Further, it is important to the cumulative effect of both temperature and rainfall on crop yields, as the effect of rainfall on crop yield is found to be positive. Studies also suggest that more than mean temperature, the changes in maximum might matter in deciding the effect of temperature change on crop yields when irrigation water is available for extended time periods.

That said, climate variability is a major concern for regions experiencing monsoon. The Indian subcontinent and the ocean surrounding it are at the center of the monsoon region (Gadgil, 2003). The Indian Monsoon has been a subject of intensive study for temporal (season and annual) variations (Gadgil, 2003). From a practical point of view, historically, many of the largest falls in crop productivity have been attributed to anomalously low precipitation events (Kumar, Kumar, Ashrit, Deshpande, & Hansen, 2004; Sivakumar, Das, & Brunini, 2005), and greater risks to food security may be posed by changes in year-to-year variability and extreme weather events (Gornall et al., 2010). However, crop model predictions do not factor in the effect of climate variability on rainfall and temperature on yield.

4.6.2.2 Impact on water productivity

Climate change drives changes in crop water productivity (which is determined by crop yield and ET) in a complex way as crop yield-related climate variables (temperature and rainfall) will have a different effect on crop evapotranspiration. While temperature can have differential impacts on cereal yield depending on the altitude, temperature increases will always increase ET. However, ET is also heavily influenced by relative humidity (inversely), which gets positively affected by precipitation occurrence. The magnitude and pattern of the same can also change the temperature of the locality with extended rainfall reducing the local temperature with a depressing effect on ET.

Carter, Hulme, and Viner (1999) note that climate change impacts on crop water productivity are affected by many uncertain factors, of which one of the most important factors is the uncertainty in global climate model predictions, especially regarding climate variability. The other factors include soil characteristics such as soil water storage (Eitzinger, Zalud, Alexandrov, et al., 2001), long-term changes in soil fertility (Sirotenko, Abashina, & Pavlova, 1997), climate variables and enhanced atmospheric CO_2 levels (Amthor, 2001), and the uncertainty of the crop growth model, which is connected with biophysical interactions. All of these factors will affect the estimation of climate change impacts on crop productivity. As long as the researchers reduce the effects of uncertain aspects, it is possible to obtain more accurate predictions about climate change impacts on crop productivity.

van de Geijn and Goudriaan (1996) found that positive climate effects on crop growth can be adjusted by effective rooting depth and nutrients; meanwhile, it can improve water productivity by 20%–40%. Howden and Jones (2004) found that changing planting dates and varieties are good measures to increase crop benefit.

Kijne, Barker, and Molden (2003) reported that water productivity can be improved by increasing investments in agricultural infrastructure and research rather than investments in the irrigation system. Water productivity concerned with water-saving irrigation is dependent on the groundwater level and evapotranspiration (Govindarajan, Ambujam, & Karunakaran, 2008). Crop water productivity is inversely related to vapor pressure (Zwart & Bastiaanssen, 2004). Crop water productivity can be increased significantly if irrigation is reduced and the crop water deficit is widely induced. In the decreased precipitation regions, the irrigation amount will increase for optimal crop growth and production, but this may decrease crop water productivity (Kijne et al., 2003).

While evapotranspiration is a major determinant of water productivity, the extent to which climate change will affect evapotranspiration and water deficits is still uncertain. Temperature, net radiation, and wind speed have positive and vapor pressure has a negative relation with PET (Goyal, 2004; Singh, 2010; Yang, Gao, Shi, Chen, & Chu, 2013). Potential evapotranspiration (PET) of wheat is significantly affected by maximum temperature ($R^2 = 0.79$), rainfall ($R^2 = 0.58$), number of rainy days ($R^2 = 0.55$), and the interactive effect of all three ($R^2 = 0.83$) (Kingra & Kukal, 2013). The authors found an increase in water productivity of wheat in central Punjab, but the same was attributed to yield improvement and not a reduction in consumptive water use.

The temperature increase is likely to enhance droughts in the future, but elevated atmospheric CO_2 concentrations, which is the cause of global warming, tend to reduce the stomatal opening in plants, which lowers the rate of transpiration from the plants. The increase in the concentration of CO_2 is likely to reduce the rate of transpiration from a few per cent for short crops to about 15% for tall crops. These reductions are, however, of comparable but opposite magnitude to predicted temperature-induced increases in evapotranspiration (Kruijt, Witte, Jacobs, & Kroon, 2008).

Guo et al. (2010) observed a positive effect of CO_2 enrichment on yield and water use efficiency. They showed that if atmospheric CO_2 increased to 600 ppm, wheat and maize yields would increase by 38% and 12% and their water use efficiencies would improve by 40% and 25%, respectively in comparison to those without CO_2 fertilization.

Xiao et al. (2007) observed a decrease in water use efficiency, length of growing period, and yield in pea-spring wheat-potato rotation with an increase in temperature and found a significant impact of supplemental irrigation in maintaining yields of crops under a climatic warming scenario.

Like in the case of crop yield, modeling studies that predict the impact of climate variability on crop water productivity are absent. Ideally, if the increase in rainfall during the growing period improves the crop yield and reduces ET (the latter due to an increase in relative humidity and lowering of temperature), then improvement in water productivity (in relation to ET) in wet years will be higher than that of normal years, and much higher than that of drought years, and vice versa. More visible will be the impact of climate variability on irrigation water productivity, as the difference in soil moisture levels will be remarkable between dry years and wet years (Kumar, 2010). Kumar (2010) found that in the two regions of the Narmada river basin where paddy was grown (Central Narmada Valley and Northern Hill Region of Chattisgarh) the irrigation water productivity varied from 1.08 and 1.74 kg/m^3, respectively in a drought year to 1.62 and 2.13 kg/m^3, respectively in a normal year.

4.7 Conclusion

The scale of measurement or assessment of water productivity changes with objectives and target audience. It can be measured at the plant level, plot/field level, farm level, system level, and basin level. With the change in scale of assessment, the determinants used in the denominator and numerator will change. For plant level measurement of water productivity, the determinants that matter are biomass output and transpiration. For plot-level assessment, it is biomass output per unit of evapotranspiration. For farm-level assessment, it is net farm income per unit of water diverted. For basin level assessment, it should be the economic value of the social, economic, and environmental benefits produced from the use of water against the total amount of water depleted which includes the total amount of water depleted in crop production plus the water that goes into the natural sink.

The methodologies for measuring/assessing water productivity vary with scale. For plant level assessment, the methodology is complex as it involves estimating ET and then partitioning it into transpiration and evaporation. Field level assessment of water productivity involves measurement of ET using lysimeters or estimation of ET using weather data, and then the measurement of crop yield. For system level and basin level assessment, several tools will have to be used in combination, including remote sensing imageries, SEBAL (soil energy balance), crop yield measurement in sample plots, along with loss of water from reservoirs and loss into the natural sink, etc. The effect of the "unit" chosen for the analysis of the results of the water productivity assessment and its implications for water management needs to be understood.

With the change in the determinants of water productivity, the mechanisms for monitoring agricultural performance, in terms of improvement in water productivity will have to be chosen carefully, in accordance with the parameters that indicate the improvement. In the case of improvement in basin-level water productivity, the determinants of water productivity are the economic value of the agricultural outputs and the total amount of water depleted for agricultural production. Since estimating the total amount of water depleted for agricultural production can be challenging, an alternative that can be adopted is to monitor the change in the annual discharge into the natural sink, the amount of water available for other sectors of the economy (industry, recreational uses) and the annual storage change in the remaining in aquifers, etc. can be indicators.

In this chapter, we also reviewed numerous scientific papers, published over the past 30 years, that dealt with the assessment of water productivity in agriculture from different parts of the world. They included assessments conducted at the plant level, plot level, and farm level. Some studies also examined the impacts of mulching and climate variability, and change on crop yield and crop water productivity. Reviews show a lack of studies on water productivity that consider irrigation systems and river levels as the unit of assessment. Reviews also show the way climate variability and change impact crop water productivity is complex and not amenable to simple formulations.

References

Aggarwal, P. K., Kalra, N., Bandyopadhyay, S. K., & Selvarajan, S. (1995). A systems approach to analyze production options for wheat in India. In J. Bouma, et al. (Eds.), Systems approaches for the design of sustainable agro-ecosystems. *Agricultural Systems*, 70(2–3), 369–393.

Ahmad, M. U. D., Bastiaanssen, W. G., & Feddes, R. A. (2002). Sustainable use of groundwater for irrigation: A numerical analysis of the subsoil water fluxes. *Irrigation and Drainage: The Journal of the International Commission on Irrigation and Drainage*, 51(3), 227–241.

Ahmad, M. U. D., Masih, I., & Turral, H. (2004). Diagnostic analysis of spatial and temporal variations in crop water productivity: A field scale analysis of the rice-wheat cropping system of Punjab. *Journal of Applied Irrigation Science, 39*(1), 43–63.

Akpalu, W., Hassan, R. M., & Ringler, C. (2008). In *Climate variability and maize yield in South Africa: Results from GME and MELE methods Environment and production technology division IFPRI discussion paper* (pp. 1–12).

Allen, R. G., Willardson, L. S., & Frederiksen, H. (1997). Water use definitions and their use for assessing the impacts of water conservation. In J. M. de Jager, L. P. Vermes, & R. Rageb (Eds.), *Sustainable irrigation in areas of water scarcity and drought* (pp. 72–81). England: Oxford.

Amthor, J. S. (2001). Effects of atmospheric CO_2 concentration on wheat yield: Review of results from experiments using various approaches to control CO_2 concentration. *Field Crops Research, 73*, 1–34.

Bastiaanssen, W., Mobin-ud-Din, A., & Zubair, T. (2003). Upscaling water productivity in irrigated agriculture using remote-sensing and GIS technologies. In *Water productivity in agriculture: Limits and opportunities for improvement* (pp. 289–300).

Burt, C. M., Howes, D. J., & Mutziger, A. (2001). Evaporation estimates for irrigated agriculture in California. In *2001 Irrigation Association Conference: San Antonio, TX*.

Carter, T. R., Hulme, M., & Viner, D. (1999). Representing uncertainty in climate change scenarios and impact studies. ECLAT-2 report no. 1. In *Proceedings of the Helsinki workshop, Norwich, UK, April 14–16* (p. 130).

Chakraborty, D., Garg, R. N., Tomar, R. K., Singh, R., Sharma, S. K., Singh, R. K., ... Kamble, K. H. (2010). Synthetic and organic mulching and nitrogen effect on winter wheat (Triticum aestivum L.) in a semi-arid environment. *Agricultural Water Management, 97*(5), 738–748.

Chakraborty, D., Nagarajan, S., Aggarwal, P., Gupta, V. K., Tomar, R. K., Garg, R. N., ... Kalra, N. (2008). Effect of mulching on soil and plant water status, and the growth and yield of wheat (Triticum aestivum L.) in a semi-arid environment. *Agricultural Water Management, 95*(12), 1323–1334.

Challinor, A. J., & Wheeler, T. R. (2008). Crop yield reduction in the tropics under climate change: Processes and uncertainties. *Agricultural and Forest Meteorology, 148*, 343–356.

Challinor, A. J., Wheeler, T. R., Craufurd, P. Q., et al. (2007). Adaptation of crops to climate change through genotypic responses to mean and extreme temperatures. *Agriculture, Ecosystem and Environment, 119*, 190–204.

Cooper, P. J. M., Keatinge, J. D. H., & Hughes, G. (1983). Crop evapotranspiration—A technique for calculation of its components by field measurements. *Field Crops Research, 7*, 299–312.

Cuculeanu, V., Tuinea, P., & Balteanu, D. (2002). Climate change impacts in Romania: Vulnerability and adaptation options. *Geology Journal, 57*, 203–209.

Dong, B., Loeve, R., Li, Y. H., Chen, C. D., Deng, L., & Molden, D. (2001). Water productivity in Zhanghe irrigation systems: Issues of scale. In R. Barker, R. Loeve, Y. H. Li, & T. P. Tuong (Eds.), *Water saving irrigation for rice*. Colombo, Sri Lanka: International Water Management Institute.

Droogers, P., van Dam, J., Hoogeveen, J., et al. (2004). Adaptation strategies to climate change to sustain food security. In J. Aerts, & P. Droogers (Eds.), *Climate change in contrasting river basins: Adaptation strategies for water, food and environment* (pp. 49–73). The Netherlands: CABI Publishing.

Eitzinger, J., Stastna, M., Zalud, Z., et al. (2003). A simulation study of the effect of soil water balance and water stress on winter wheat production under different climate change scenarios. *Agric Water Management, 61*, 195–217.

Eitzinger, J., Zalud, Z., Alexandrov, V., et al. (2001). A local simulation study on the impact of climate change on winter wheat production in northeastern Austria. *Ecological Economics, 52*, 199–212.

Feddes, R. A., Hoff, H., Bruen, M., Dawson, T., De Rosnay, P., Dirmeyer, P., ... Pitman, A. J. (2001). Modeling root water uptake in hydrological and climate models. *Bulletin of the American Meteorological Society, 82*(12), 2797–2810.

Frederiksen, H. D., & Allen, R. G. (2011). A common basis for analysis, evaluation and comparison of off stream water uses. *Water International, 36*, 266–282. https://doi.org/10.1080/02508060.2011.580449.

Frederiksen, H. D., Allen, R. G., Burt, C. M., & Perry, C. (2012). Responses to Gleick et al., which was itself a response to Frederiksen and Allen. *Water International, 37*(2), 183–197.

Fuchs, M., & Hadas, A. (2011). Mulch resistance to water vapor transport. *Agricultural Water Management, 98*(6), 990–998.

Gadgil, S. (2003). The Indian monsoon and its variability. *The Annual Review of Earth and Planetary Sciences, 31*, 429–467.

Gahukar, R. T. (2009). Food security: The challenges of climate change and bioenergy. *Current Science, 96*(1), 26–28.

van de Geijn, S. C., & Goudriaan, J. (1996). The effects of elevated CO2 and temperature change on transpiration and crop water use. In F. Bazzaz, & W. G. Sombroek (Eds.), *Global climate change and agriculture production* (pp. 1–21). Chichester, UK: Wiley.

Gichuki, F., Cook, S., & Turral, H. (2006). In *Agricultural water productivity: issues, concepts and approaches Basin focal project working paper 1 (no. 618-2016-41183)*.

Gleick, P. H., Christian-Smith, J., & Cooley, H. (2011). Water use efficiency and productivity rethinking the basin approach. *Water International, 36*, 784–798. https://doi.org/10.1080/02508060.2011.631873.

Gornall, J., Betts, R., Burke, E., Clark, R., Camp, J., Willett, K., & Wiltshire, A. (2010). Implications of climate change for agricultural productivity in the early twenty-first century. *Philosophical Transactions of the Royal Society B: Biological Sciences, 365*(1554), 2973–2989.

Govindarajan, S., Ambujam, N. K., & Karunakaran, K. (2008). Estimation of paddy water productivity (WP) using hydrological model: An experimental study. *Paddy Water Environment, 6*, 327–339.

Goyal, R. K. (2004). Sensitivity of evapotranspiration to global warming: A case study of arid zone of Rajasthan (India). *Agricultural Water Management, 69*, 1–11.

Guo, R., Lin, Z., Mo, X., & Yang, C. (2010). Responses of crop yield and water use efficiency to climate change in the North China plain. *Agricultural Water Management, 97*, 1185–1194.

Howden, M., & Jones, R. N. (2004). Risk assessment of climate change impacts on Australia's wheat industry. In *New directions for a diverse planet: Proceedings of the 4th international crop science congress, Brisbane, Australia, Sept. 26–Oct. 1* (p. 1848).

Hundal, S. S., & Kaur, P. (2007). Climate change and its impact on cereal productivity in Indian Punjab: A simulation study. *Current Science, 92*(4), 506–511.

Hussain, I., Sakthivadivel, R., Amarasinghe, U., Mudasser, M., & Molden, D. (2003). *Land and water productivity of wheat in the western Indo-Gangetic plains of India and Pakistan: A comparative analysis. Vol. 65*. IWMI.

Jabran, K., Ullah, E., Hussain, M., Farooq, M., Zaman, U., Yaseen, M., & Chauhan, B. S. (2015). Mulching improves water productivity, yield and quality of fine rice under water-saving rice production systems. *Journal of Agronomy and Crop Science, 201*(5), 389–400.

Kang, Y., Khan, S., & Ma, Y. (2009). Climate change impacts on crop yield, crop water productivity and food security–A review. *Progress in Natural Science, 19*(2009), 1665–1674.

Kendy, E., Molden, D. J., Steenhuis, T. S., Liu, C., & Wang, J. (2003). *Policies drain the North China Plain: Agricultural policy and groundwater depletion in Luancheng County, 1949–2000. (Vol. 71)*. IWMI.

Kere, G. M., Nyanjage, M. O., Liu, M., & Nyalala, S. P. O. (2003). Influence of drip irrigation schedule and mulching materials on yield and quality of greenhouse tomato (Lycopersicon esculentum mill 'money maker'). *Asian Journal of Plant Sciences*, 2(14), 1052–1058.

Kijne, J. W., Barker, R., & Molden, D. (2003). Improving water productivity in agriculture: Editors' overview. In J. W. Kijne, R. Barker, & D. Molden (Eds.), *Water productivity in agriculture: Limits and opportunities for improvement* (pp. xi–xix). Wallingford UK: CABI Publishing.

Kingra, P. K., & Kukal, S. S. (2013). Impact of climate change on evapotranspiration and water productivity of wheat in Central Punjab. *Journal of Research Punjab Agricultural University*, *50*, 181–183.

Krishnan, P., Swain, D. K., Bhaskar, B. C., et al. (2007). Impact of elevated CO2 and temperature on rice yield and methods of adaptation as evaluated by crop simulation studies. *Agricultural Ecosystem and Environment*, *122*, 233–242.

Kruijt, B., Witte, J. M., Jacobs, C. M. J., & Kroon, T. (2008). Effects of rising atmospheric CO_2 on evapotranspiration and soil moisture: A practical approach for the Netherlands. *Journal of Hydrology*, *349*, 257–267.

Kumar, M. D. (2010). *Managing water in river basins: Hydrology, economics, and institutions*. New Delhi: Oxford University Press.

Kumar, M. D. (2014). *Changes in water productivity in agriculture with particular reference to South Asia: Past and the future*. New Delhi: Final report submitted to International Water Management Institute.

Kumar, K. K., Kumar, K. R., Ashrit, R. G., Deshpande, N. R., & Hansen, J. W. (2004). Climate impacts on Indian agriculture. *International Journal of Climatololgy*, *24*, 1375–1393.

Kumar, M. D., Scott, C. A., & Singh, O. P. (2011). Inducing the shift from flat-rate or free agricultural power to metered supply: Implications for groundwater depletion and power sector viability in India. *Journal of Hydrology*, *409*(1–2), 382–394.

Kumar, M. D., Turral, H., Sharma, B., Amarasinghe, U., & Singh, O. P. (2008). Water saving and yield enhancing micro-irrigation technologies in India: When and where can they become best bet technologies? In *Proceedings of the 7th Annual Partners' Meet of IWMI-Tata water policy research program*. International Water Management Institute, South Asia Regional Office, ICRISAT Campus, Hyderabad, 2–4 April 2008.

Kumar, M. D., & van Dam, J. C. (2009). Improving water productivity in agriculture in India: Beyond 'More Crop per Drop'. In Kumar, M. D., & Amarasinghe, U. (Eds.), *Water productivity improvements in Indian agriculture: Potentials, constraints and prospects. Strategic Analyses of the National River Linking Project (NRLP) of India, Series 4, 163*.

Kumar, M. D., & van Dam, J. C. (2013). Drivers of change in agricultural water productivity and its improvement at basin scale in developing economies. *Water International*, *38*(3), 312–325.

Lal, R. (2005). Climate change, soil carbon dynamics, and global food security. In: R. Lal, B. Stewart, N. Uphoff, et al. (Eds.), *Climate change and global food security* (pp. 113–143). Boca Raton, FL: CRC Press.

Leblanc, M., Tweed, S., Van Dijk, A., & Timbal, B. (2012). A review of historic and future hydrological changes in the Murray-Darling basin. *Global and Planetary Change*, *80*, 226–246.

Loeve, R., Dong, B., Molden, D., Li, Y. H., Chen, C. D., & Wang, J. Z. (2004). Issues of scale in water productivity in the Zhanghe irrigation system: Implications for irrigation in the basin context. *Paddy and Water Environment*, *2*, 227–236.

Luo, Q., Williams, M., Bellotti, W., et al. (2003). Quantitative and visual assessments of climate change impacts on south Australian wheat production. *Agricultural Systems*, *77*, 173–186.

Masumoto, T., Hai, P. T., & Shimizu, K. (2008). Impact of paddy irrigation levels on floods and water use in the Mekong River basin. *Hydrological Processes: An International Journal*, 22(9), 1321–1328.

Molden, D., Murray-Rust, H., Sakthivadivel, R., & Makin, I. (2003). A water-productivity framework for understanding and action. In J. W. Kijne, R. Barker, & D. J. Molden (Eds.), *Vol. 1. Water productivity in agriculture: limits and opportunities for improvement* CABI Publishing.

Molden, D., Oweis, T. Y., Pasquale, S., Kijne, J. W., Hanjra, M. A., Bindraban, P. S., ... Hachum, A. (2007). *Pathways for increasing agricultural water productivity, H040200*. International Water Management Institute.

Molle, F., Mamanpoush, A., & Miranzadeh, M. (2004). *Robbing Yadullah's water to irrigate Saeid's garden: Hydrology and water rights in a village of central Iran, IWMI Research Report # 80*. Colombo, Sri Lanka: International Water Management Institute.

Ngangom, B., Das, A., Lal, R., Idapuganti, R. G., Layek, J., Basavaraj, S., ... Ghosh, P. K. (2020). Double mulching improves soil properties and productivity of maize-based cropping system in eastern Indian Himalayas. *International Soil and Water Conservation Research*, 8(3), 308–320.

Ortiz, R., Sayre, K.D., & Govaerts, B. (2008). Climate change: Can wheat beat the heat? *Agriculture, Ecosystem and Environment*, 126(1–2), 46–58.

Palanisami, K., Senthilvel, S., Ranganathan, C. R., & Ramesh, T. (2006). Water productivity at different scales under tank, Canal and Well Irrigation. In *Centre for agricultural and rural development studies* Tamil Nadu Agricultural University, Coimbatore.

Parry, M., Rosenzweig, C., Iglesias, A., et al. (1999). Climate change and world food security: A new assessment. *Global Environmental Change*, 9, S51–S67.

Perry, C. (2007). Efficient irrigation; inefficient communication; flawed recommendations. *Irrigation and Drainage: The Journal of the International Commission on Irrigation and Drainage*, 56(4), 367–378.

Perry, M. W., & d'Antuono, M. F. (1989). Yield improvement and associated characteristics of some Australian spring wheat cultivars introduced between 1860 and 1982. *Australian Journal of Agricultural Research*, 40(3), 457–472.

Popova, Z., & Kercheva, M. (2005). CERES model application for increasing preparedness to climate variability in agricultural planning-risk analyses. *Physics and Chemistry of the Earth*, 30, 117–124.

Reddy, V. R., & Pachepsky, Y. A. (2000). Predicting crop yields under climate change conditions from monthly GCM weather projections. *Environmental Modelling & Software*, 15(1), 79–86.

Schmidhalter, U., & Oertli, J. J. (1991). Transpiration/biomass ratio for carrots as affected by salinity, nutrient supply and soil aeration. *Plant and Soil*, 135, 125–132.

Seckler, D., Molden, D. J., & Sakthivadivel, R. (2003). The concept of efficiency in water resources management and policy. In J. W. Kijne, R. Barker, & D. J. Molden (Eds.), *Vol. 1. Water productivity in agriculture: Limits and opportunities for improvement* CABI Publishing.

Siddique, K. H. M., Belford, R. K., Perry, M. W., & Tennant, D. (1989). Growth, development and light interception of old and modern wheat cultivars in a Mediterranean-type environment. *Australian Journal of Agricultural Research*, 40(3), 473–487.

Siddique, K. H. M., Kirby, E. J. M., & Perry, M. W. (1989). Ear: Stem ratio in old and modern wheat varieties; relationship with improvement in number of grains per ear and yield. *Field Crops Research*, 21(1), 59–78.

Siddique, K. H. M., Tennant, D., Perry, M. W., & Belford, R. K. (1990). Water use and water use efficiency of old and modern wheat cultivars in a Mediterranean-type environment. *Australian Journal of Agricultural Research*, 41(3), 431–447.

Singh, R. (2005). *Water productivity analysis from field to regional scale: Integration of crop and soil modeling, remote sensing and geographic information.* Doctoral thesis The Netherlands: Wageningen University. Wageningen.

Singh, R. K. (2010). Sensitivity of evapotranspiration to climate change: A case study of Jaipur district in Rajasthan. *The Asian Man - An International Journal, 4*(1), 39–48.

Singh, B., Eberbach, P. L., Humphreys, E., & Kukal, S. S. (2011). The effect of rice straw mulch on evapotranspiration, transpiration and soil evaporation of irrigated wheat in Punjab, India. *Agricultural Water Management, 98,* 1847–1855.

Singh, R., van Dam, J. C., & Feddes, R. A. (2006). Water productivity analysis of irrigated crops in Sirsa district, India. *Agricultural Water Management, 82,* 253–278.

Singh, V. P., Singh, P. K., & Bhatt, L. (2014). Use of plastic mulch for enhancing water productivity of off-season vegetables in terraced land in Chamoli district of Uttarakhand, India. Land capability classification and land resources planning using remote sensing and GIS 3. *Journal of Soil Water Conservation, 13*(1), 68–72.

Sirotenko, O. D., Abashina, H. V., & Pavlova, V. N. (1997). Sensitivity of Russian agriculture to changes in climate, CO2 and tropospheric ozone concentrations and soil fertility. *Climate Change, 36,* 217–232.

Sivakumar, M. V. K., Das, H. P., & Brunini, O. (2005). Impacts of present and future climate variability and change on agriculture and forestry in the arid and semi-arid tropics. *Climate Change, 70,* 31–72. https://doi.org/10.1007/s10584-005-5937-9.

Tanner, C. B., & Sinclair, T. R. (1983). Efficient water use in crop production: research or re-search? In *Limitations to efficient water use in crop production* (pp. 1–27).

Tojo Soler, C. M., Sentelhas, P. C., & Hoogenboom, G. (2007). Application of the CSM-CERES-maize model for planting date evaluation and yield forecasting for maize grown off-season in a subtropical environment. *European Journal of Agronomy, 27,* 165–177.

UNDESA. (2016). Basin wide groundwater management using the system of nature: Kumamoto city, Japan. Water for life, UN-Water Best Practices Award 2013 edition: Winners, www.un.org/waterforlife decade/winners2013.shtml (accessed 2 September 2016).

Walker, N. J., & Schulze, R. E. (2006). An assessment of sustainable maize production under different management and climate scenarios for smallholder agro-ecosystems in KwaZulu-Natal, South Africa. *Physics and Chemistry of the Earth, 31,* 995–1002.

Wang, Z., Jin, M., Šimůnek, J., & van Genuchten, M. T. (2014). Evaluation of mulched drip irrigation for cotton in arid Northwest China. *Irrigation Science, 32*(1), 15–27.

Wang, Z., Wu, Q., Fan, B., Zheng, X., Zhang, J., Li, W., & Guo, L. (2019). Effects of mulching biodegradable films under drip irrigation on soil hydrothermal conditions and cotton (Gossypium hirsutum L.) yield. *Agricultural Water Management, 213,* 477–485.

Xiao, G., Zhang, Q., Yao, Y., Yang, S., Wang, R., Xiong, Y., & Sun, Z. (2007). Effects of temperature increase on water use and crop yields in a pea-spring wheat-potato rotation. *Agricultural Water Management, 91,* 86–91.

Xie, Z., Wang, Y., Jiang, W., & Wei, X. (2006). Evaporation and evapotranspiration in a watermelon field mulched with gravel of different sizes in Northwest China. *Agricultural Water Management, 81*(1–2), 173–184.

Xie, Z. K., Wang, Y. J., & Li, F. M. (2005). Effect of plastic mulching on soil water use and spring wheat yield in arid region of Northwest China. *Agricultural Water Management, 75*(1), 71–83.

Yang, X., Gao, W., Shi, Q., Chen, F., & Chu, Q. (2013). Impact of climate change on the water requirement of summer maize in the Huang-Huai-Hai farming region. *Agricultural Water Management, 124,* 20–27.

Yao, F., Xu, Y., Lin, E., et al. (2007). Assessing the impacts of climate change on rice yields in the main rice areas of China. *Climate Change, 80,* 395–409.

Yuan, C., Lei, T., Mao, L., Liu, H., & Wu, Y. (2009). Soil surface evaporation processes under mulches of different sized gravel. *Catena, 78*(2), 117–121.

Zhang, P., Wei, T., Han, Q., Ren, X., & Jia, Z. (2020). Effects of different film mulching methods on soil water productivity and maize yield in a semiarid area of China. *Agricultural Water Management, 241*. November 01, 2020.

Zhang, F., Zhang, W., Qi, J., & Li, F. M. (2018). A regional evaluation of plastic film mulching for improving crop yields on the loess plateau of China. *Agricultural and Forest Meteorology, 248*, 458–468.

Zhu, Z., Giordano, M., Cai, X., & Molden, D.J. (2004). The Yellow River Basin: Water accounting, water accounts, and current issues. *Water International, 29*(1), 2–10.

Zwart, S. J., & Bastiaanssen, W. G. M. (2004). Review of measured crop water productivity values for irrigated wheat, rice, cotton and maize. *Agricultural Water Management, 69*, 115–133.

CHAPTER 5

Past trends in water productivity at the global and regional scale

5.1 Introduction

At the global scale, opportunity for improving agricultural water productivity exist through the following routes: (i) growing crops in regions that offer advantage in terms of high solar energy, low aridity (Zwart & Bastiaanssen, 2004) and good soils; (ii) improving the utilization of the available water in the basins which otherwise goes uncaptured and flow into natural sinks, for crop production either by expansion of rainfed crops or by diverting the developed water resources for irrigated crop production (Molden et al., 2007); (iii) by improving the efficiency of water use at the level of irrigation schemes through improvement in the efficiency of utilization of water from reservoirs (by reducing reservoir evaporation), and reduction in nonrecoverable seepage losses in the conveyance; (iv) improving the efficiency of use of water at the farm level by improving the allocative efficiency of water through diverting more water for economically more water-efficient crops (Gichuki, Cook, & Turral, 2006; Kumar, 2005); (v) improving the efficiency of use of water at the field/plot level by reducing the evaporation/ ET ratio plus reducing the nonrecoverable deep percolation (Siddique, Tennant, Perry, & Belford, 1990); and (vi) improving the efficiency of use of water by the plants by improving the transpiration coefficient and raising the harvest index (Schmidhalter & Oertli, 1991).

Having said that, the availability of renewable water resources in relation to the arable land, and per capita arable land availability, the two most important factors driving agricultural production potential, vary across countries. Further, different countries have followed different trajectories of the development of water resources, especially for irrigation. While many countries in the semiarid tropics show a very high degree of exploitation of water resources, many in Africa have utilized a small fraction of their

Current Directions in Water Scarcity Research, Volume 3
https://doi.org/10.1016/B978-0-323-91277-8.00003-4

utilizable water resources (Alexandratos & Bruinsma, 2012; FAO, 2011). On the other hand, many countries in Europe and the Americas having temperate climates withdraw very little water for irrigation (Kumar, Malla, & Tripathy, 2008), all having significant implications for agricultural outputs and crop water productivity.

Also, different agricultural regions of the world have different climatic conditions—from tropical wet (found in Central and South America as well as Africa and southwest Asia) to tropical wet & dry (found in South and Central America, Africa, and parts of Asia) and dry to desert climate (Arabian desert, Sahara desert) to semiarid to the Mediterranean (on the coast of the Mediterranean sea, US west coast to Australia) to humid subtropical (southeastern part of the United States and large parts of China) to the marine west coast, humid continental to Tundra (in the northern hemisphere around the arctic ocean) to highland to Subarctic. Climate has a strong effect on crop water productivity with different weather parameters (viz., solar radiation, temperature, relative humidity, and wind speed) having differential effects on evapotranspiration, and temperature and solar radiation having an effect on yield (Aggarwal, Kalra, Bandyopadhyay, & Selvarajan, 1995). Further, the extent of use of agricultural technologies including crop technologies also varies from country to country, depending on the level of development of life sciences and agricultural technologies, economic condition, the agricultural growth needs, and the usefulness of these technologies for raising productivity.

All these factors have influenced the past growth trends in agricultural water productivity. This chapter reviews the past growth trends in agricultural water productivity globally and regionally, taking a 40-year time period. Since it is a humongous task to analyze the changes in crop water productivity at various levels (plant, field, farm, irrigation system, and basin and country-level) owing to the difficulty in obtaining data for the relevant indicators of such changes, the analysis is confined to certain broad indicators of the resultant changes at the macrolevel such as rate of change in yield of major crops, rate of change in the value of agricultural outputs and change in total water withdrawal for agriculture.

5.2 Global trends in agricultural water productivity

Even today, agriculture is the largest consumer of water in the world, among all the water consuming sectors. Within agriculture, food production requires the largest amount of water. Globally, the demand for food has been growing during the past nearly five decades, more because of population growth—recorded at 1.69% per annum, and less due to the increase in per capita consumption. The growth in global demand for agricultural commodities in calorific terms has been very modest—from 2373 kcal/person/ day in 1969–1971 to 2772 kcal/person/day in 2005/2007. This change has been mostly attributed to changes in consumption patterns, with a shift toward meat and dairy products. While the direct cereal consumption went up from 144 kg/person/year to 160 kg/person/year, the quantum of cereals for all uses went up from 304 kg/person/annum to 330 kg/person/annum.[a] Globally, the growth in consumption of many food items, such as milk and sugar has been marginal, while that of pulses and roots has been negative. The consumption of other food items, in calorific terms, increased from 194 to 325 kcal/person/day (source: Alexandratos & Bruinsma, 2012).

5.2.1 Historical growth in food consumption at the regional level

But the regional trends are distinctly different from global trends. The historical changes in per capita food consumption, region-wise is given in Table 5.1. It shows that not only the per capita consumption against the developed countries, the growth in per capita consumption has also been very minimal, except for East Asia. The only notable exceptions are in the case of milk and milk products in South Asia—from a lowest of 42 kg/capita/year in 1979—81 to 71 kg in 2005/07, and food cereals in the case of Near East/North Africa. In the case of Near East/North Africa, the consumption of food cereals has been consistently higher than that of developed countries since 1979–1981. In East Asia, the consumption levels for all food items have been growing exponentially. In the case of meat, milk and vegetable oils, there has been a quantum jump in consumption levels (source: based on table 2.6, Alexandratos & Bruinsma, 2012). This was mainly driven by the changes in consumption behavior in China, owing to high growth in average income and hundreds of millions of people being lifted out of poverty during the past three decades.

---

[a] The additional cereal demand is affected by increased cereal demand for feeding poultry and other livestock, raised for milk and meat. The consumption of meat also grew impressively from 26 kg/person/annum to 49 kg/person per annum, mostly due to an increased preference for livestock-based diet in the developed countries

Table 5.1 Change in consumption of food in South Asia (1979–1971 to 2005–2007) in kilogram per annum.

Type of food	Sub-Saharan Africa			Near East/North Africa			Latin America and the Caribbean			South Asia			East Asia			Developed countries		
	1979–81	1989–91	2005–07	1979–81	1989–91	2005–07	1979–81	1989–91	2005–07	1979–81	1989–91	2005–07	1979–81	1989–91	2005–07	1979–81	1989–91	2005–07
Cereals	126	115	125	205	213.0	203	129.0	130.0	138.0	147.0	161	152	160	172	163	156	162	167
Roots and tubers	174	184	184	27	32.0	37.0	73.0	64.0	63.0	20.0	18.0	25.0	82	61	60	84	78	77
Sugar and sugar crops	9.5	8.20	10.7	27.7	28.1	27.80	47.0	44.0	42.0	19.40	20.6	19.1	8.3	10.6	13.4	40	36	34
Dry pulses	9.8	9.00	10.50	6.2	7.90	7.10	12.60	10.50	11.30	11.10	12.0	10.3	4.3	2.6	1.9	2.9	2.90	2.90
Vegetable oils, oil seeds, and products	8.1	8.20	9.40	10.3	12.3	12.30	10.10	12.0	13.60	5.70	7.10	8.90	4.8	7.6	9.9	14	16	19
Meat	10.2	10.20	10.1	16.60	19.0	23.70	40.0	42.0	61.0	4.00	4.60	4.40	13.4	22.8	44.3	74	80	80
Milk and dairy (excluding butter)	33.0	29.0	31.0	83.0	74.0	78.0	96.0	94.0	111.0	42.00	54.00	71.0	5.0	7.4	23.3	195	201	202
Other food (kcal/person/day)	n/a	119	126	252	306.0	343.0	250.0	251.0	264.0	87.00	98.00	137.0	140	202	367	508	498	458

Source: Alexandratos, N., & Bruinsma, J. (2012). World agriculture towards 2030/2050: the 2012 revision.

5.2.2 Historical growth in global agricultural production and driving factors

In order to meet the growing demand, the production of agricultural commodities increased at a rate of 2.2% per annum for nearly four and a half decades from 1961/63 to 005/07. During the period from 1980 to 2012, the value of agricultural production (in 2004–06 prices) in the world increased at a rate of 2.25% per annum from US $ 1180 billion to US $ 2411 (Fig. 5.1).

'It is important to look at the impact of increased food production on water resources. As per the World Water Development Report 2 of the United Nations, globally, water withdrawals tripled in 50 years when the population doubled (UN Water, 2006). But agricultural water withdrawals grew at a much lower rate, increasing by roughly 50% (source: authors' own estimates based on Shiklomanov, 2000 and FAO AQUASTAT, 2012) in spite of having a modest growth in average calorie intake and high population growth. This has been possible through an almost 100% increase in water productivity in agriculture between 1961 and 2001. The major factor behind this achievement has been yield growth (see Table 5.2), mostly resulting from irrigation and the adoption of modern crop technologies. Globally, while 80% of the cropland, which is rain-fed, produces 60% of the agricultural production, the remaining 20% of irrigated land produces the rest 40% (UN Water, 2006). This means that irrigated crops have at least 2.7 times more productivity (yield per ha of land) than rain-fed crops.

For many crops, the yield increase has occurred without increased water consumption, and sometimes with even less water, owing to improvements in the harvest index. Two crops for which water consumption experienced little variation during these years are rice (mostly irrigated) and wheat

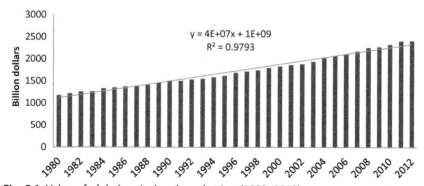

Fig. 5.1 Value of global agricultural production (1980–2012).

Table 5.2 Relative contribution of various factors to growth in crop production in the world.

Regions of the world	Source of growth in crop production in the World during 1961–2007		
	Arable land expansion (%)	Increase in cropping intensity (%)	Yield growth (%)
Sub-Saharan Africa	31	31	38
Near East and North Africa	17	22	62
Latin America and the Caribbean	40	7	53
South Asia	6	12	82
East Asia	28	−6	77
World	14	9	77

Source: Bruinsma, J. (2011), "The resource outlook to 2050: By how much do land, water use and crop yields need to increase by 2050?". Chapter 6. In P. Conforti, (Ed.), Looking ahead in World Food and Agriculture: Perspectives to 2050. Rome: FAO http://www.fao.org/docrep/014/i2280e/i2280e06.pdf.

(mostly rain-fed), for which the recorded increases in yield globally amount to 100% and 160%, respectively (FAO, 2003a). This happened through the introduction of fast-growing and high-yielding varieties.

At the global level, since 1960, the increase in water consumption for agriculture has been 800 km^3 (Shiklomanov, 2000) while the world population has doubled to 6 billion. Considering that the rain-fed cropland has not increased in a major way on the global scale, it can be safely concluded that with an additional 800 km^3 of water the world has been able to feed an additional 3 billion people. This gives a rough estimate of 0.72 m^3 per day (of water to produce food for a person). This figure is low compared to the estimated global average for 2000 of 2.4 m^3 of water per day per person, which includes water for food at the field level, not including water losses. At the global level, the average water needs for food per capita stood at around 3 m^3 per day, including the losses, which was just half of the average water needs of 1961 (i.e., 6 m^3 per day per person) (Renault, 2003).

As can be seen from Table 5.2, globally, growth in crop production during 1961–2007 was driven by yield growth, which contributed 77% to the total increase in crop production during 1961–2007, and arable land and cropping intensity contributing only 14% and 9%, respectively. But there are sharp regional differences in the relative contribution of these factors. For instance, in Sub-Saharan Africa, the growth in crop production was driven by expansion in cropped land and cropping intensity, contributing to an extent of 62%, whereas the contribution of yield growth was only 38%.

In Sub-Saharan Africa, the yield of cereals had stagnated since 1980—at around 1 ton per ha—, which was a marginal improvement from around 0.60 ton per ha in the early 1960s (Hazell & Wood, 2008). Contrary to this, in South Asia, the growth in production mostly came from yield growth (82%). In South Asia, the average yield of cereals went up from 1.40 ton per ha in 1980 to 2.5 ton per ha in 2005 (Hazell & Wood, 2008).

Table 5.3 shows the growth in yield of three major crops in different regions of the world, during three different time periods, i.e., 1980–1989, 1990–1999, and 2000–2007 and the current yields. The figures show that the current yields for the three crops in developed countries are far higher than in all other regions. The only exception is wheat, wherein the average yield in East Asia is far greater than that of developed countries (Hengsdijk & Langeveld, 2009). Here again, the very low average yields obtained in the US bring down the average yield values for "developed countries," in spite of the fact that many European countries have very high yields, and in some instances even higher than the climatic potential (Fischer, van Velthuizen, & Nachtergaele, 2009). But, interestingly, as Table 5.3 shows, during the 27-year period from 1980 to 2007, developed countries recorded much lower growth in yield as compared to the developing countries, while they started with high average yields in the base year. Lower yield growth in Sub-Saharan Africa means faster growth in the demand for arable land to meet the growing demand for food.

Before we begin to examine how this has driven changes in crop water productivity, it is important to understand the implications of yield growth on crop water productivity. Yield growth can come from three sources mainly. First, raising the genetic yield potential of the crop like the case of semi-dwarf variety of wheat. Second, from raising the bottom of the yield distribution by increasing resistance to stress (Mann, 1999). Third, by applying optimum dosage of nutrients inputs and irrigation water, and soil amelioration as required for the plant growth. Yield growth through the first two routes can change water productivity in kg per ET. In the case of the third option, in what way the crop water productivity at the field level would change depends on how the evapotranspiration from the field has changed in relation to the yield response curve of ET.[b] As regards fertilizers, increased

---

[b] At very low levels of ET, improvements in ET would increase crop water productivity as well as yield, while at higher levels, the WP would decline sharply, though yield would increase, and then level off.

Table 5.3 Current yields and average growth rates in yield of three major crops in different regions of the world.

	Current average yield (ton per ha)			Annual growth in wheat yield (%)			Annual growth in maize yield (%)			Annual growth in rice yield (%)		
	Wheat	Maize	Rice	1980–89	1990–99	2000–07	1980–89	1990–99	2000–07	1980–89	1990–99	2000–07
D. Countries	3.79	6.75	11.5	1.65	1.37	0.09	1.45	0.2	1.5			
S. S. Africa		1.49	1.62				1.35	0.45	1.2	1.5	0.5	−0.4
North Africa	2.25			2.4	2.8	2						
Latin America	2.44	3.095		2.4	2	1.5	2.25	2.2	2.7			
South Asia	2.51	2.37	3.35	2.5	2.3	1.7	1.4	2.1	2.1	2	2	1.5
East Asia	4.48	5.22	6.26	4.3	2.4	1.8	3.3	2.4	0.4	2.5	1.5	1

Source: Based on figs. 3.2, 3.3, and 3.4 of Hengsdijk and Langeveld (2009): Yield Trends and Yield Gap Analysis of Major Crops in the World, Work Document 170, Wageningen University and Research Centre, December. The values of growth rates for developed countries, and Sub-Saharan Africa and Latin America were obtained by taking the average of the corresponding figures for the subregions, which belong to these regions/countries, for which data were available in the report.

nutrient inputs would increase crop water productivity as well as yield, as more water from the soil gets converted into beneficial transpiration.

An extensive review of studies in developed and developing countries concluded with convincing evidence of the increasing genetic yield potential of wheat (Rejesus, Heisey, & Smale, 1999). The larger proportion of genetic gains in yield of wheat, rice, and maize also results from increased stress tolerance rather than gains in yield potential per se. Much of the progress in developing stress tolerance in wheat has come from dramatically improved resistance to wheat diseases, particularly the rusts (Sayre, Singh, Huerta-Espino, & Rajaram, 1998). But there is ample evidence that yield advances in farmers' wheat fields have slowed worldwide. Some of this slowdown may be partially explained by reduced growth in demand for wheat, but it is noteworthy that yield increases in advanced wheat-producing areas of developing countries (e.g., north-western Mexico and the Indian and Pakistan Punjab) have also slowed in the past 10–15 years. The reasons for slower growth in yield gains in farmers' fields are many and complex; it is quite likely that crop management issues and resource degradation play important.

To analyze the implications of growth in crop production on water productivity, let us first take the example of Sub-Saharan Africa. The irrigation water withdrawal is very low in Sub-Saharan Africa (Fig. 5.2), a total of 55 BCM (Bruinsma, 2011), with a total of 5.6 m ha of cropland equipped to receive irrigation water. But expansion in arable land has been phenomenal. Intuitively, at the regional level, this has led to water productivity

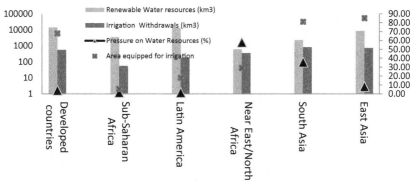

Fig. 5.2 Water withdrawal for irrigation, different regions. *Source: Bruinsma, J. (2011), "The resource outlook to 2050: By how much do land, water use and crop yields need to increase by 2050?". Chapter 6. In P. Conforti, (Ed.), Looking ahead in World Food and Agriculture: Perspectives to 2050. Rome: FAO http://www.fao.org/docrep/014/i2280e/i2280e06.pdf.*

improvements due to the following: increased utilization of water naturally available in the soil profile for crop production, by converting nonbeneficial evaporation from land into beneficial transpiration, through arable land expansion; the crops, which were earlier rain-fed, receiving some supplementary irrigation resulting in enhanced yields; and some increase in cropping intensity with the increased availability of water for irrigation, through small and medium water resource development projects.

In South Asia, which has the highest cropping intensity among all regions of the world (Siebert, Portmann, & Döll, 2010), irrigation water withdrawals had increased remarkably, which is very high at present (35.6%), with a total volumetric withdrawal of 817 BCM for a total cropped area of 81 m ha equipped to receive irrigation water (Bruinsma, 2011).

During the period from 2001 to 2010, the irrigated area in four of the South Asian countries (India, Nepal, Pakistan, and Afghanistan) increased from 79.26 to 86.38 m per ha (source: FAO AQUASTAT for the corresponding years). Its impact on crop water productivity both at the farm level and regional level has been phenomenal, as rain-fed, monsoon crops were replaced by irrigated varieties and new irrigated crops are grown in the winter and summer seasons with water available from new storages and diversions systems created in the region. The drivers of change in crop water productivity in the region are utilization of stream-flows—which otherwise goes uncaptured as drainage outflows—, and tapping of groundwater for irrigation; and effective utilization of residual soil moisture in the land after the harvest of monsoon crop, for raising winter crops, and to a lesser extent the utilization of some amount of available arable land for crop production— by converting non-beneficial evaporation into beneficial transpiration.

Unlike Sub-Saharan Africa and South Asia, in Latin America, increased crop production (from 1961 to 2007) was achieved mainly through yield growth (52%), followed by expansion in arable land (40%) (see Table 5.2). Intuitively, the water, which is drawn from the natural system, is being diverted for raising crops in the land which were earlier not cultivated, or for providing supplementary irrigation to the existing rain-fed crops, with the latter contributing to yield growth. Along with this, a lot of the land under forests is converted into cropland (Ramankutty et al., 2018). The very high precipitation received in most parts of Latin America reduces the irrigation water requirements. Here again, one could infer that water productivity at the regional level had improved as a result of the following: conversion of beneficial evapotranspiration in natural vegetation into beneficial transpiration for crop production, through arable land expansion; and, effective utilization of blue water (irrigation water) for increasing crop ET and thereby crop yields.

In East Asia, production growth (during 1961–2007) was mainly achieved through substantial improvements in yield of major crops and arable land expansion, while cropping intensity witnessed some reduction (see Tables 5.2 and 5.3). In China, yield growth was achieved by spreading the available irrigation water in a larger area and making efficient use of rainwater. In the 5 years, 7.7 m per ha of irrigation areas were improved with water-saving techniques nationally in China, with which, the national irrigation area meeting the water-saving standard reached 16.7 m per ha, accounting for 31% of the effective irrigation area. The area adopting water-injection seeding technique, walking irrigation machinery, and controlled irrigation for paddy fields, and other non-engineering facilities reached 16.7 m ha; the irrigated water for a unit area on the national average dropped from 7140 m^3 per ha in 1995 to 6585 m^3 per ha in 2000, and the water consumption for 10,000 Yuan worth of agricultural production dropped from 1917 m^3 in 1996 to 1591 m^3 in 2000. Under the condition that the total irrigation water cannot be increased, the potential is tapped in water-saving measures. In this way, the newly-added 4.3 m ha of irrigation area was provided with water, and 6.7 m ha of farmlands found its water source for dry seeding. But for the above water-saving measures, the agricultural water demand would have been increased by 25 BCM according to the original water usage (Yuping, 2001).

5.2.3 Crop water productivity at field scale for different regions

To understand the full implications of agricultural growth driven by crop yield (kg per ha) on water use, it is important to understand how crop water productivity varies between regions. Crop water productivity is a function of ET, nutrient inputs, climate, and soil characteristics and therefore can change from location to location for the same variety of crop and can also change with variety (source: based on Molden, Murray-Rust, Sakthivadivel, & Makin, 2003; Abdullaev & Molden, 2004; Schmidhalter & Oertli, 1991; Siddique et al., 1990; Zwart & Bastiaanssen, 2004). The review of research on crop water productivity, in relation to the total amount of water depleted, undertaken by Zwart and Bastiaanssen (2004) for four major crops across the globe showed sharp variations in values of CWP across regions, countries, and locations. They are summarized in Table 5.4. Table 5.4 also provides explanations in situations where maximum CWP was observed.

Table 5.4 Measured values of crop water productivity in four major crops in different countries.

Name of crop	No. of countries considered	Water productivity range (kg per m³)	Location for highest CWP: CWP (kg per m³)	Yield and ET values for maximum CWP case	Reasons for high CWP
Wheat	13	0.60–1.70	China: 2.67	7150 kg per ha and 268 mm	Application of manure, mulching leading to higher production and lower ET
Rice	8	0.60–1.60	China: 2.20	10,000 kg per ha and 465 mm	Alternate wetting and drying of rice plots
Maize	10	1.10–2.70	China: 3.99	9058 kg per ha and 226 mm	Alternate furrow and deficit irrigation; reduced ET to 226 mm
Cotton Lint	9	0.14–0.33	China and Israel: 0.35	2000 kg of lint yield, and 617 mm	Cotton was planted in furrows, with plastic mulching, leaving holes near the plants. This reduced soil evaporation and improved soil water

Note: Only between 5% and 95% percentile points were considered for determining the CWP range.
Source: Based on Zwart, S.J., Bastiaanssen, W.G.M. (2004). Review of measured crop water productivity values for irrigated wheat, rice, cotton and maize. *Agricultural Water Management, 69,* 115–33.

It is evident from the mapping of crop yield in relation to ET that the yield response to ET is not a linear relationship and that high water productivity values could be obtained by manipulating ET to be in a range where the yield response is high, and also by maximizing the transpiration part of ET and minimizing soil evaporation part, notwithstanding the influence of solar radiation and relative humidity on CWP. In this regard, the study by Zwart and Bastiaanssen (2004) found that the regions within the latitude of 30°–40° had the highest water productivity.

5.2.4 Crop water productivity at the regional level

Now let us analyze the implications of growth in crop outputs on water productivity at the regional level. If we make a realistic assumption that no water is imported into or exported from any of the regions under consideration, and that only the available water resources (both green water and blue water) in the basins of these regions are tapped, and that there is no large-scale mining of groundwater in these regions for agriculture), the growth rates in crop outputs at the regional level can be treated as growth rates in crop water productivity. The underlying assumption here is that the denominator of regional water productivity, i.e., the total amount of water depleted, which includes the water which goes into the natural sinks, evaporation outflows, and the water beneficially used in crop production as ET, hasn't changed over the given time period. The growth rates for crop productivity in different regions of the world are given in Table 5.5.

As can be seen from the table, the highest growth in crop production and therefore regional crop water productivity during 1987–2007 was

Table 5.5 Historical growth in crop production in different regions of the world.

Region	Annual percentage growth during		
	1961–2007	1987–2007	1997–2007
World	2.20	2.30	2.30
Sub-Saharan Africa	2.60	3.30	3.00
Latin America and the Caribbean	2.70	2.90	3.70
Near East/North Africa	2.90	2.50	2.40
South Asia	2.60	2.40	2.10
East Asia	3.40	3.60	3.20
Developed countries	0.80	0.40	0.50

Source: Alexandratos, N., & Bruinsma, J. (2012). World agriculture towards 2030/2050: the 2012 revision.

experienced in East Asia (3.6%), followed by Sub-Saharan Africa (3.3%). This can be explained by the very high growth in yield obtained by China in some important crops such as maize, wheat, and paddy. The wheat yield in East Asia increased from around 2000 kg per ha in 1980 to around 4800 kg per ha in 2005/2006 (source: based on fig. 9 in Bruinsma, 2009). Instead of increasing cropping intensity using the new sources created for irrigation, the strategy in China has been to spread out the available water to provide critical irrigation to crops that were earlier rain-fed and even expand the arable land, thereby utilizing the water in the soil profile (see Table 5.2). The very high measured values of water productivity for four major crops in China are explained by this. Contrary to this, in Sub-Saharan Africa, owing to poor yields, an increasing amount of arable land was put to rain-fed cultivation. Hence, the growth in crop productivity was driven by extensive use of water in the soil profile, rather than irrigation. Hence, at the field level, water productivity is very low for these rain-fed crops (Brauman, Siebert, & Folly, 2013; Rockström, Barron, & Fox, 2002). In a global water productivity mapping covering 13 major crops, Brauman et al. (2013) found that the estimated crop water productivity for maize was lowest in large parts of SS Africa (0.3 kg per m^3), while it was highest (1.7 kg per m^3) in the US corn belt and north China plains.

South Asia had one of the lowest growths in crop production (therefore, crop water productivity at the regional level) during the last two decades (2.4%), next only to the group of developed countries. In the preceding two decades, both South Asia and Sub-Saharan Africa had the second-lowest growth rates (2.6%), after the developed countries. Yet these growth rates are remarkable. But this remarkable growth in crop water productivity regional level cannot be confused with the same degree of improvement in water productivity at the farm level, or even basin level. It is quite likely that an increasing quantum of water from the basins of the region is being diverted for crop production through large and small water development projects and small water harvesting systems, along with measures that drive field level improvement in water productivity, including the selection of improved varieties, changes in cropping pattern, irrigation technology changes, and improved fertilizer usage. This has been the case in South Asia.

Globally, crop water productivity has been increasing almost steadily during 1987–2007.

Interestingly, if we strictly go by the crop production growth as an indicator of growth in water productivity, the growth in regional crop water productivity appears to be very low in developed countries, despite the high level of adoption of efficient irrigation technologies such as drips and sprin-klers, and relatively higher crop yields for major crops, viz., wheat, rice, and maize.

There are many reasons for this. First, the arable land under crop produc-tion is reducing in many of these developed countries (Bruinsma, 2009), with growing pressure to release more water for the environment. The sub-stantial increments in average yields achieved for many crops, which are far higher than the developing countries,[c] and the very modest increase in food demand[d] had made this possible. In other cases, an increasing amount of water stored in reservoirs is being diverted for urban uses, manufacturing, and recreation. Thus, while the water productivity at the farm level is increasing, much less water from the natural system is being used in crop production, though this partial analysis provides a distorted picture. In reality, in most developed countries, the overall water productivity in terms of $ output per unit of water has been sharply increasing from 1980 onward, as a result of the use of efficient water use technologies in the domestic sector, and efficient production systems in the manufacturing sector (Margat & Andréassian, 2008).

More than growth in crop production from the available water resources in a region or subregion, the overall growth in the agricultural sector is what would ultimately determine the water productivity gains. This depends on how the crops are used for producing agricultural product, which is high in the value chain. The crops (like green fodder or dry fodder obtained from crops such as paddy, wheat, pearl millet, sorghum, and groundnut) produced on the farm can be diverted to feeding dairy animals or animals raised for meat. Similarly, the grains produced on the farm can be diverted for feeding livestock, including poultry and animals for meat. The grains produced in the farm or the small fish (fish stock) can also be used to feed the fish stock in fisheries. Hence, it is important to look at the growth in agricultural

---

[c] The average wheat yields in major wheat-producing, developed countries of Europe viz., Germany, UK, France, Holland and Denmark are higher than in South Asia and Sub-Saharan Africa (Bruinsma, 2009); the average maize yield in North America is far higher than in Asia, Latin America and Africa (FAO, 2003b); the average rice yields in Oceania is the highest in the world (11.658 t per ha), against 3.96 t per ha in Asia, and 2.23 t per ha in Africa (Kubo & Purevdorj, 2004).

[d] In the developed countries, per capita food consumption increased marginally from 3288 kcal/capita/day in 1989/1991 to 3360 kcal/capita/day in 2005/2007. Whereas the developed country population increased marginally from 1079 million in 1970 to 1351 million in 2006, recording a growth rate of 0.6% per annum (Alexandratos & Bruinsma, 2012).

productivity. Such farming systems can change water productivity at the farm level (Amede, Tarawali, & Peden, 2011; Kumar & van Dam, 2013). The growth rates in agricultural production in different regions of the world are presented in Table 5.6.

It is evident from Table 5.6 that East Asia had the highest growth rates in agricultural production during 1987–2007, and therefore the improvement in water productivity is also expected to be the highest during this period for this region. This was followed by SS Africa, and then South Asia. Interestingly, in spite of recording a high growth rate in crop production (3.50% per annum), the growth rate in overall agricultural production obtained, in value terms, was only 3.20% in Sub-Saharan Africa. Whereas in South Asia, the growth rate in agricultural output was 2.70% against a growth rate of 2.4% in crop output. During the period from 1980 to 2012, the annual growth in the value of agricultural outputs in South Asia was 3.15%, with a quantum jump in the output from $ 126 billion to $ 340 billion (source: authors' own estimates based on FAO AQUASTAT for the respective years).

This means, in the South Asian region, the value addition in agriculture through subsidiary activities such as tree production, fisheries, milk production, poultry farming, and livestock production is increasingly growing over a period of time, whereas in SS Africa, this value addition is increasingly becoming less over time. A lot of this can be attributed to composite farming systems with crop and livestock in South Asian countries (Baltenweck et al., 2003), wherein crop resides (from groundnut) and by-products of cereals as dry fodder are used for animal feeding, and animal dung is extensively used as

Table 5.6 Historical growth in agricultural production in value terms in different regions of the world.

Region	Annual percentage growth during		
	1961–2007	1987–2007	1997–2007
World	2.20	2.20	2.20
Sub-Saharan Africa	2.60	3.20	3.10
Latin America and the Caribbean	2.90	3.30	3.80
Near East/North Africa	3.00	2.70	2.60
South Asia	2.90	2.70	2.40
East Asia	4.00	4.20	3.30
Developed countries	0.90	0.20	0.50

Source: Alexandratos, N., & Bruinsma, J. (2012). World agriculture towards 2030/2050: the 2012 revision.

farmyard manure to improve soil fertility. Studies show that there could be significant gains in water productivity at the farm level if livestock production uses crop by-products as major inputs (Amede et al., 2011; Kumar et al., 2008; Kumar & van Dam, 2013). The highest value addition in agriculture is occurring in East Asia, with a 4.2% growth in agricultural output against 3.60% growth in crop outputs. This is because of the large-scale production of livestock products, which are higher up in the value chain. One reason for the low growth in value of agricultural outputs in SS Africa is that the domestic demand for agricultural produce, at the higher end of the value chain such as fish and milk, have not been growing at the rate, experienced in East and South Asia (Bruinsma, 2009).[e]

There can be issues associated with drawing inferences on agricultural water productivity trends based on trends in agricultural growth if such growth is driven by large-scale imports of inputs. For instance, as noted by Singh, Sharma, Singh, and Shah (2004), milk production can be dependent on the import of cattle feed and fodder. Similarly, rearing of livestock for meat can also be dependent on the import of animal feed—which is based on cereals. Estimating the real water productivity should take into account the water embedded (virtual water) in feed and fodder (Singh et al., 2004). While the region, under consideration, would be gaining in terms of water, what is important is that the total amount of water depleted in producing the agricultural output in that region would be higher than the total amount of internal water resources of the region which is depleted for production. Hence, the actual water productivity in agriculture would be lower than the figures obtained through "accounting" for the region's water resources.

5.3 Summary

The annual growth in crop production globally (2.2%–2.3%) has been mainly attributed to yield growth (77%), followed by arable land expansion (14%), and an increase in cropping intensity (9%). But there is remarkable variation in the extent of contribution of different factors across regions. Generally, in the developed countries, the modest growth in agricultural

---

[e] While in East Asia and South Asia the demand for milk and milk products grew at an annual rate of 6.7 and 4.3 respectively, in Sub-Saharan Africa, it grew at a much lower rate of 2.3%. Whereas in the case of meat, the demand grew at 2.7% per annum in SS Africa, while the growth rate in East Asia and South Asia were 6.4% and 2.1%, respectively (Bruinsma, 2009).

output (0.5% during 1997–2007) is mainly from a substantial increase in crop yields, while the areas under crop cultivation and cropping intensity have actually declined. In developing countries, while yield gain has contributed majorly to crop output growth, the contribution of intensified cropping and arable land expansion is greater. The contribution of arable land expansion and increase in cropping intensity is very high (31% for each) in Sub-Saharan Africa, whereas in the case of South Asia, the contribution of increased cropping intensity, which is made possible through irrigation expansion, is significant (12%). In South Asia, irrigation water withdrawal in relation to total renewable water availability is excessively high, given the monsoon climate and the limited scope for arable land expansion. It is next only to near-East/North Africa. In Sub-Saharan Africa, irrigation water withdrawal is still very low.

The overall economic value of agricultural outputs has increased the world over during 1961–2007, with an annual compounded growth rate of 2.2%. The growth rate was only marginally higher than the world average in South Asia (2.9%). In Sub-Saharan Africa, the growth rate was even lower (2.6%). In East Asia, the growth rate during the period was an impressive 4.0%. In developing countries, the growth rate was 3.3% during 1961–2007. Given the historical trends in crop outputs and the main factor that has contributed to the output growth (i.e., yield growth), the value of agricultural outputs, and irrigation water withdrawals, it can be inferred that water productivity in crop production has increased globally and more so in developing countries. The highest improvement in water productivity, however, is expected to have occurred in East Asia, which had recorded a remarkable and consistent increase in crop yields without a substantial increase in irrigation water withdrawal. Though the increase in crop outputs, the extent of contribution of yield growth to this growth, and the increase in the value of agricultural outputs suggest increasing water productivity in South Asia over time, the rate of improvement (in physical and economic productivity of water in crop production) is unlikely to be as high as that of crop outputs or value of these outputs, respectively, as a quantum jump in irrigation water withdrawals in these countries (especially India, Pakistan, Bangladesh, and Iran) might also have contributed to yield growth.

Another interesting trend is that over the years, the growth rate in agricultural outputs and value of these outputs had declined in South Asia, whereas it has increased in Sub-Saharan Africa. One reason for this trend observed in South Asia is the decline in yield growth of major crops.

References

Abdullaev, I., & Molden, D. (2004). Spatial and temporal variability of water productivity in the Syr Darya Basin, Central Asia. *Water Resources Research, 40*(8).

Aggarwal, P. K., Kalra, N., Bandyopadhyay, S. K., & Selvarajan, S. (1995). A systems approach to analyze production options for wheat in India. In J. Bouma *et al.* (eds.) Systems approaches for the design of sustainable agro-ecosystems. *Agricultural Systems, 70*(2–3), 369–393.

Alexandratos, N., & Bruinsma, J. (2012). *World agriculture towards 2030/2050: The 2012 revision.*

Amede, T., Tarawali, S., & Peden, D. (2011). Improving water productivity in crop livestock system of drought-prone regions, editorial comment. *Exploratory Agriculture, 47*(S1), 1–5.

Baltenweck, I., Staal, S., Ibrahim, M. N. M., Hererro, M., Holman, F., Jabbar, M. A., ... Waithaka, M. (2003). *Crop-livestock intensification and interaction across three continents (No. 610-2016-40501).*

Brauman, K. A., Siebert, S., & Folly, J. A. (2013). Improvements in crop water productivity increase water sustainability and food security—A global analysis. *Environment Research Letters, 8,* 204030.

Bruinsma, J. (2009). The resource outlook to 2050: By how much do land, water and crop yields need to increase by 2050. In *Vol. 2050. Expert meeting on how to feed the world* (pp. 24–26).

Bruinsma, J. (2011). The resource outlook to 2050: By how much do land, water use and crop yields need to increase by 2050? Chapter 6 In P. Conforti (Ed.), *Looking ahead in world food and agriculture: Perspectives to 2050.* Rome: FAO. http://www.fao.org/docrep/014/i2280e/i2280e06.pdf.

Fischer, G., van Velthuizen, H., & Nachtergaele, F. (2009). *Global agro-ecological assessment-the 2009 revision.*

FAO AQUASTAT. (2012). *Datasets from FAO's Global Information System on Water and Agriculture for the year 2012.* Food and Agriculture Organization of the United Nations. http://www.fao.org/aquastat/en.

Food and Agriculture Organization of the United Nations. (2003a). Unlocking the water potential of agriculture, Rome. Accessed from http://www.fao.org/3/y4525e/y4525e00.htm#Contents.

Food and Agriculture Organization of the United Nations (2003b). Maize Post Harvest Operations, INPhO Post Harvest Compendium, FAO, Rome.

Food and Agriculture Organization of the United Nations. (2011). *The state of World's land and water resources for food and agriculture: Managing systems at risk.* Rome and Routledge, London: FAO.

Gichuki, F., Cook, S., & Turral, H. (2006). Agricultural water productivity: issues, concepts and approaches. In *Basin focal project working paper 1 (no. 618-2016-41183).*

Hazell, P., & Wood, S. (2008). Drivers of change in global agriculture. *Philosophical Transactions of the Royal Society B: Biological Sciences, 363*(1491), 495–515.

Hengsdijk, H., & Langeveld, J. W. A. (2009). *Yield trends and yield gap analysis of major crops in the world (no. 170).* Wettelijke Onderzoekstaken Natuur & Milieu.

Kubo, M., & Purevdorj, M. (2004). The future of rice production and consumption. *Journal of Food Distribution Research, 35*(1), 1–15.

Kumar, M. D. (2005). Impact of electricity prices and volumetric water allocation on energy and groundwater demand management: Analysis from Western India. *Energy Policy, 33*(1), 39–51.

Kumar, M. D., Malla, A. K., & Tripathy, S. (2008). Economic value of water in agriculture: Comparative analysis of a water-scarce and a water-rich region in India. *Water International, 33*(2), 214–230.

Kumar, M. D., & van Dam, J. C. (2013). Drivers of change in agricultural water productivity and its improvement at basin scale in developing economies. *Water International, 38*(3), 312–325.

Mann, C. C. (1999). *Crop scientists seek a new revolution.*

Margat, J., & Andréassian, V. (2008). *L'eau. Vol. 155.* Le Cavalier Bleu.

Molden, D., Murray-Rust, H., Sakthivadivel, R., & Makin, I. (2003). A water-productivity framework for understanding and action. In J. W. Kijne, R. Barker, & D. J. Molden (Eds.), *Vol. 1. Water productivity in agriculture: Limits and opportunities for improvement* CABI.

Molden, D., Oweis, T. Y., Pasquale, S., Kijne, J. W., Hanjra, M. A., Bindraban, P. S., ... Hachum, A. (2007). *Pathways for increasing agricultural water productivity, H040200.* International Water Management Institute.

Ramankutty, N., Mehrabi, Z., Waha, K., Jarvis, L., Kremen, C., Herrero, M., & Rieseberg, L. H. (2018). Trends in global agricultural land use: Implications for environmental health and food security. *Annual Review of Plant Biology, 69*, 14.1–14.27.

Rejesus, R. M., Heisey, P. W., & Smale, M. (1999). *Sources of productivity growth in wheat: A review of recent performance and medium-to long-term prospects.* CIMMYT.

Renault, D. (2003). Value of virtual water in food: principles & virtues. In A. Hoekstra (Eds.), *Proceedings of the expert meeting held 12–13 December 2002, Delft, The Netherlands.* UNESCO-IHE.

Rockström, J., Barron, J., & Fox, P. (2002). Rainwater management for increased productivity among small-holder farmers in drought prone environments. *Physics and Chemistry of the Earth, Parts A/B/C, 27*(11–22), 949–959.

Sayre, K. D., Singh, R. P., Huerta-Espino, J., & Rajaram, S. (1998). Genetic progress in reducing losses to leaf rust in CIMMYT-derived Mexican spring wheat cultivars. *Crop Science, 38*(3), 654–659.

Schmidhalter, U., & Oertli, J. J. (1991). Transpiration/biomass ratio for carrots as affected by salinity, nutrient supply and soil aeration. *Plant and Soil, 135*(1), 125–132.

Shiklomanov, I. A. (2000). World water resources and water use: Present assessment and outlook for 2025. In F. Rijsberman (Ed.), *World water scenarios: Analysis.* The Hague, WWF2.

Siddique, K. H. M., Tennant, D., Perry, M. W., & Belford, R. K. (1990). Water use and water use efficiency of old and modern wheat cultivars in a Mediterranean-type environment. *Australian Journal of Agricultural Research, 41*(3), 431–447.

Siebert, S., Portmann, F. T., & Döll, P. (2010). Global patterns of cropland use intensity. *Remote Sensing, 2*(7), 1625–1643.

Singh, O. P., Sharma, A., Singh, R., & Shah, T. (2004). Virtual water trade in dairy economy. *Economic and Political Weekly, 39*(31).

UN Water. (2006). Water a shared responsibility. UN World Water Development Report 2006. UN Water, New York, March 16, 2006.

Yuping, G. (2001). *Water-Saving Irrigation Practice in China--Demands, Technical System, Current Situation, Development Objective, and Countermeasures.*

Zwart, S. J., & Bastiaanssen, W. G. M. (2004). Review of measured crop water productivity values for irrigated wheat, rice, cotton and maize. *Agricultural Water Management, 69*, 115–133.

CHAPTER 6

Implications of future growth in demand for agricultural commodities and climate change on land and water use and water productivity

6.1 Introduction

The future needs to improve water productivity in agriculture would be driven by the likely growth in demand for food in the coming decades, and the resource availability constraints—particularly the availability of land and water (Kumar, Bassi, & Singh, 2020). The global projections show that the world will be able to meet the increased demand for agricultural commodities through increased production not only at the aggregate level but also at the regional level (Alexandratos & Bruinsma, 2012). However, there are two important issues that need to be addressed. First, the growth in food demand in developing countries with low per capita calorie intake will be mainly driven by the growth in per capita income, and the present low per capita calorie intake is due to the high incidence of poverty with hundreds of millions of people not having sufficient money to buy food. According to the 2019 global hunger index assessment, which considered 117 countries, several developing countries of Africa and South Asia are food insecure, with their values for global hunger index being very high (source: https://www.globalhungerindex.org/results.html#country-level-data). Millions of children in these countries suffer from malnutrition and high infant mortality. Therefore, from the point of view of food and nutritional security, a rate of growth in agricultural production that equals the demand growth is simply insufficient. Second, even with the modest level of growth in production to meet the projected future demand, the pressure on water and land resources, and the implications for agricultural water productivity need to be examined.

Such pressures will depend on how a country is placed with respect to the relative availability of water and land. In countries with abundant arable land and water resources, such pressures are very unlikely. In countries having plenty of arable lands but extremely limited fresh-water resources, pressure on water resources is likely to increase with increased demand for food and other agricultural commodities, and water productivity improvement in agriculture will be quite essential (Kumar et al., 2020). However, if crops can be grown in such land without irrigation (meaning, sufficient amount of green water is available), then simple expansion of cultivated area with proper management of rainwater could contribute to an increase in agricultural outputs, and this strategy can reduce the pressure on blue water (Rosa, Chiarelli, Rulli, Angelo, & D'Odorico, 2020). In regions having extremely limited land resources, but plenty of water when compared with the size of the land that can be irrigated, agricultural output per unit of arable land will have to be increased. This requires not only higher crop yields, but also higher biomass yield per day. That said, as we have seen in Chapter 2, the situation with regard to per capita arable land availability and availability of utilizable water resources (in terms of water adequacy ratio) varies across the world. In certain countries, even when there is sufficient water available for harnessing in the future for crop production, the cost of irrigation development can be extremely high. In such cases also, there will be a need to focus on water productivity.

Climate change will further complicate the future scenario with respect to the demand for land and water for meeting the global demand for agricultural commodities, because of the impact it can have on crop yields and crop water productivity (Kang, Khan, & Ma, 2009). In this chapter, we will review the situation in different regions of the world with regard to the future possibilities for raising crop yields, expanding arable land, increasing irrigation potential or intensifying cropping, and improving water productivity for meeting agricultural growth challenges. Finally, drawing on the research studies reviewed in Chapter 4 vis-à-vis climate change impacts on crop yields and crop water productivity, the constraints, and opportunities offered by likely changes in climate to raising crop yields, improving crop water productivity, and increasing agricultural outputs will also be analyzed.

6.2 Projected future growth in demand for agricultural commodities at the global level

Alexandratos and Bruinsma (2012) in a study, which looked at World Agriculture in 2030/50 done for FAO notes that significant parts of the world population will reach per capita consumption levels that do not leave much scope for further increases and that there would be negative growth rates of aggregate food demand in countries where per capita consumption levels are or will be high as their population starts declining in the later part of the projection period. Most developed countries have largely completed the transition to livestock-based diets, while not all developing countries—for instance India—are likely to shift in the foreseeable future to levels of meat consumption typical of western diets. Thus, the growing needs of world food production to meet the growth of demand will be lower than in the past, even after accounting for increases in per capita consumption and changes in diets.

FAO (2006a) estimates that the growth in global food demand will progressively decline from 2.2% a year in the last decades of the 20th century, to 1.6% in 2015, 1.4% in 2015–2030, and 0.9% in 2030–2050. Part of the current and future pressure on water resources comes from increasing demands for animal feed. Meat production requires 8–10 times more water than cereal production (FAO, 2006a). As per the latest projections available, there will be an average increase of 0.6% a year in irrigated land until 2030, compared with 1.6% over 1971–2003 (source: based on FAOSTAT, 2007). Because of continued increases in water productivity in agriculture, FAO (2006b) estimates that 36% more food will be produced with 13% more water.

Considering the main regions, of particular interest is the extent to which the two regions with low and largely inadequate per capita food consumption levels, viz., Sub-Saharan Africa and South Asia, may progress to higher levels. South Asia's level is not different from that of 10 or 20 years ago, while Sub-Saharan Africa has made some, progress. South Asia's average is heavily weighted with India, which, is characterized by a lack of improvement in per capita food consumption in kcal/person/day (Alexandratos & Bruinsma, 2012). Long-term projections by Alexandratos and Bruinsma (2012) show that both regions would break with the history of no, or sluggish, improvement, and by 2050 may reach levels near those that the other three developing regions have at present.

Table 6.1 Projections of demand for agricultural commodities in different regions of the world.

Region	Projected growth in demand of agricultural commodities		
	2005/2007–2030	2030–2050	2005/2007–2050
World	1.40	0.80	1.10
Sub-Saharan Africa	2.60	2.10	2.40
North East/North Africa	1.70	1.10	1.50
Latin America and the Caribbean's	1.70	0.60	1.20
South Asia	2.00	1.30	1.70
East Asia	1.40	0.50	1.00
Developed countries	0.60	0.20	0.50

Note: while dealing with heterogeneous commodities, such as cereals, meat, milk, and nonfood products, the international dollar prices of the commodities were used as weightage for each commodity to arrive at the growth rates.
Source: Alexandratos, N., & Bruinsma, J. (2012). World agriculture towards 2030/2050: the 2012 revision.

The projections for global demand for (all) agricultural commodities are given in Table 6.1. It shows the marked difference in demand growth rates between developed countries and developing countries, beyond 2005/2007 till 2050. The total cereal demand is projected to increase to 3009 m ton from 2068 m ton in 2005/2007, with that for animal feed accounting for a little more than 50% of the total. The developing countries would add 740 m ton of the total demand growth of 940 m ton that is likely to occur in 4 and a half decades. Among developing countries, the highest growth in food demand is expected to come from Sub-Saharan Africa, followed by South Asia. The primary driver of this demand growth is the population growth in these regions; the secondary driver being the changes in consumption levels and pattern—with greater intake of high-value food such as animal products (meat and milk), sugar, vegetable oil, and fruits & vegetables, replacing cereals. The expected changes in consumption of food commodities (in kcal/person/day), which drive this growth in different regions, are provided in Fig. 6.1 (source: Alexandratos & Bruinsma, 2012).

The trend clearly shows that South Asia and Sub-Saharan Africa would experience a major rise in calorie intake in the next 30 years. Yet, their consumption levels would be far less than the current consumption levels of developed countries. But East Asia would reach consumption levels that are almost at par with that of developed countries, owing to a substantial rise in consumption of meat—from 44.3 to 71.1 kg per person per year and

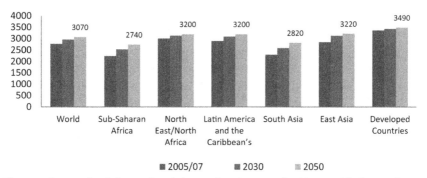

■ 2005/07   ■ 2030   ■ 2050

Fig. 6.1 Current food demand and demand projections (kcal/person/day), 2030/2050.

vegetable oil consumption going up from 9.9 to 15.3 kg per person per year, attributed to the growing affluence of the region.

At the global level, future production of agricultural commodities would balance the demand for food. However, certain regions could experience some deficits in the production of food commodities as their growth in production may not keep pace with the demand growth, owing to a scarcity of land, water, and other factors of production. Such deficits will have to be met through trade (Kumar, 2014). As shown by Alexandratos and Bruinsma (2012), growth rates in food production in the developing regions have been a little below those of consumption, and as a result, their agricultural imports have been growing faster than their exports, leading to the gradual erosion of their traditional status as agricultural trade surplus group. The surplus has diminished and turned into a net deficit since the late 1970s. Without Brazil, which has emerged as a major agricultural commodity-exporting country, the net trade deficit of all developing countries continued to increase rapidly till 2005/2007. In the future, these trends may be attenuated, given that the slowdown in the growth of demand will lead to the net import needs of the developing countries growing slower than in the past. Nevertheless, the developing countries will have to increase the production of food commodities in the future to meet the higher growth in food demand, as compared with developed countries.

What impact this will have on water productivity in agriculture in these countries would be decided by a variety of complex factors. One of them is how much additional arable land area could be put to crop production in these countries, without the use of irrigation water, and the yield can be obtained under rain-fed conditions. If much of the additional production required in the future comes from rain-fed areas without any adverse impact on renewable water resources that are being tapped for other sectors of the water economy, this can have a positive impact on crop productivity at the regional level. This depends on how much uncultivated arable land is available in these countries, and where they are located (Kumar, 2014).

It is estimated that a total of 960 m ha of land is available from different developing regions in the world, which is suitable for rain-fed farming. But its distribution is very uneven. Nearly 85% is located in Sub-Saharan Africa and Latin America. Other developing countries, especially South Asia, have almost no extra arable land (Alexandratos & Bruinsma, 2012).

Another factor is the yield growth different regions are likely to achieve for major crops without much impact on crop transpirative demand (T), and the amount of land area suitable for growing such crops these regions have. Higher crop yields obtained through plant genetic engineering would directly add to crop water productivity at the field level and also at the basin/regional level. The third factor is the technical efficiency in the use of irrigation water required to intensify cropping. The technical efficiency improvement would reduce nonbeneficial consumptive uses. Since this would reduce irrigation water withdrawal requirements, this also will have a positive impact on agricultural water productivity both at the field level and basin/regional level. The fourth factor is the scope for enhancing crop yield through better on-farm water management, including management of rainwater. Many of these variables are hard to predict and are also intertwined. But, the least developed countries of SS Africa are least likely to make significant progress on these fronts (Kumar, 2014).

In sum, growth in demand for food in the future will have to be compensated by an equal increase in productivity of water used for irrigation for raising the basic agricultural commodities used in their production, unless additional land is available for either expanding arable land or water resources for increasing cropping intensity or enhancing yield. In the next section, we will examine how and where the additional agricultural output is likely to be generated.

6.3 Projected future growth in agricultural outputs at the global level and its implications for agricultural water productivity

In the earlier section, we have seen how the demand for agricultural commodities is likely to grow in different regions of the world. It is important to know the likely trajectory of growth in agricultural production in those

Table 6.2 Projections of future agricultural outputs in different regions of the world.

Region	Projected growth in production of agriculture commodities		
	2005/2007–2030	2030–2050	2005/2007–2050
World	1.30	0.80	1.10
Sub-Saharan Africa	2.50	2.10	2.30
North East/North Africa	1.60	1.20	1.40
Latin America and the Caribbean's	1.70	0.80	1.30
South Asia	1.90	1.30	1.60
East Asia			
Developed countries	0.70	0.30	0.50

Note: while dealing with heterogeneous commodities, such as cereals, meat, milk, and nonfood products, the international dollar prices of the commodities were used as weightage for each commodity to arrive at the growth rates.
Source: Alexandratos, N., & Bruinsma, J. (2012). World agriculture towards 2030/2050: the 2012 revision.

regions, to assess the implications of the said demand growth in terms of pressure on water resources.

Based on the past trends in arable land expansion, base year data of yield of major crops, past growth trends in yields of major crops under both rain-fed and irrigated conditions, expansion in irrigation facilities, and the constraints induced by the availability of arable land in different regions, Alexandratos and Bruinsma (2012) had forecasted the potential future production of agricultural commodities (Table 6.2), future growth in yield of major crops and accordingly worked out the additional land areas that are likely to be brought under farming in these regions (Table 6.3). It is clear from Table 6.3 that expansion in arable land (0.1% per annum) won't achieve much growth in food production to match with the demand growth (1.0% over the 35-year period), additional growth has to come from irrigation expansion, water use efficiency improvements and yield growth, which is not related to irrigation improvements.

While in SS Africa, arable land expansion is expected to be 0.44% per annum (Table 6.3), this would directly contribute to production growth and regional water productivity improvements, as soil water would be tapped for crop production, even if this will not have any impact on field and farm level water productivity. But, the slight expansion in irrigation in the region would also result in an enhanced yield of rain-fed crops, thereby enhancing crop water productivity at the field level (Rockström, Barron, & Fox, 2002).

Table 6.3 Factors driving agricultural water productivity in different regions of the world.

Region name	Arable land suitable for cultivation (m per ha)	Arable land in use (m per ha)	Projected growth in arable land use (%)	Projected growth in irrigation (%)	Pressure on water resources (%)
World	4495	1592	0.10	0.10	7.20[*]
Sub-Saharan Africa	1073	240	0.44	0.50	3.80
Latin America and the Caribbean's	1095	202	0.49	0.30	1.60
Near East/ North Africa	95	84	0.00	0.20	54.10
South Asia	195	206	0.08	0.10	40.0–42.0[*]
East Asia	410	236	0.00	0.20	10.0–11.0[*]
Developed countries	1592	624	−0.14	0.00	4.00

Source: Based on data presented in various output tables of Alexandratos and Bruinsma (2012) and [*]authors' own estimates.

We need to reckon with the fact that even as per 2009 estimates, SS Africa had less than 25% of its cultivable land under crop production (FAO, 2009). The region has the lowest irrigated to rain-fed area ratio of less than 3% (FAO, 2006a, 2006b, fig. 5.2, pp. 177). As a result, the drought-prone areas of Sub-Saharan Africa are characterized by one of the lowest levels of agricultural productivity in the world. One of the reasons for the poor utilization of arable land for cultivation is the lack of assurance in obtaining yields, in the wake of uncertain rainfalls and the absence of irrigation facilities.

Irrigation is the key to improving water security, expanding cropland, and also raising crop yields in drought-prone areas. But one of the biggest challenges in reducing the region's vulnerability to droughts and associated problems of food insecurity and hunger is in developing this irrigation (FAO, 2007, 2009). Currently, only two of the Sub-Saharan African countries have extensive irrigation. They are South Africa and Madagascar, with 1.43 million ha and 1.15 million ha of irrigated area, respectively (You, Wood, & Sichra, 2007).

In the case of India and Pakistan, two south Asian countries with extensive irrigation, the initial impetus in irrigation development came from the public sector, mainly covering large medium and minor surface irrigation systems in India and large Indus Basin Irrigation System in Pakistan (FAO, 2012). But later on with the advancement in drilling technology, massive rural electrification, institutional financial for well development, and heavy subsidies for electricity for agricultural use, groundwater development for irrigation took place in a big way in the countryside, with well irrigation becoming intensive in the semiarid and arid parts of both the countries (Kumar, 2007; Qureshi, McCornick, Sarwar, & Sharma, 2010).

But irrigation development trajectory is unlikely to be the same in Sub-Saharan Africa, given the drastically different sociopolitical scenario and human resource capacities of the countries of that region. Surface irrigation development in Sub-Saharan Africa is likely to happen at a slow pace even in the coming years and too little can be done about changing it from a water sector perspective, unless the macrolevel issues of political instability, corruption in governments, institutional capacity, and finance are addressed to. These are in addition to the host of social and environmental issues that the surface water projects raise. The corruption in government is likely to reduce the donor confidence in many of the countries there (Kumar, Sivamohan, & Narayanamoorthy, 2012).

Public sector spending remains the dominant source of finance for irrigation (Foster & Briceño-Garmendia, 2010). While the spending needs for developing and sustaining irrigation schemes in the continent were estimated at US $3.4bn annually (Foster & Briceño-Garmendia, 2010), the average actual spending in the 2001–2006 period was the only US $0.9bn (Briceño-Garmendia, Smits, & Foster, 2008). Hence there is a notable funding gap between what is needed and what is forthcoming. The overall scenario of the agriculture sector has been no different. The public spending on agriculture has been less than 10% in 40 out of the 43 SS African countries for which data were available. More importantly, public spending on agriculture, as a percentage of the total government spending, has been less than its share of the country's GDP in all SS African countries (Goyal & Nash, 2016).

The main leverage, therefore, lies in groundwater development through private sector initiatives. But, unlike India, where drilling technology has come very handy in rural areas due to low cost and easy accessibility, well drilling is very expensive in most African countries (Namara, Gebregziabher, Giordano, & De Fraiture, 2013). Apart from issues of the high cost of well development due to high drilling costs, there are also issues of uncertainty about resource conditions (Foster, Tuinhof, & Garduño, 2006;

Kumar et al., 2012).[a] Unless farmers are convinced about their ability to obtain water, investments are unlikely to come, even if funds are available. An associated challenge is rural electrification (Kumar et al., 2012). The other challenge is to change the investment climate. Apart from irrigation, there is a need for greater investment in public goods that support agriculture, such as research and extension, rural roads, storage facilities, education, and health (FAO, 2009).

After Cai and Rosegrant (2003), water productivity of irrigated cereal crops is expected to be much higher than that of rain-fed crops—to the tune of 0.15–0.20 kg/m^3 for paddy and 0.10–0.40 kg/m^3 for other cereals. Irrigation expansion will also lead to a much higher production outcome owing to an increase in cropping intensity. Since only a small fraction of the region's water resources are currently exploited (Kumar et al., 2012), irrigation expansion through water resources development would further add to water productivity improvements at the regional level also, as the water which went untapped would be harnessed. However, it is important to note that there are big institutional constraints for irrigation development affected by the lack of adequate capacity of the agencies concerned to plan, design, build, and operate water infrastructure projects, and the poor financial health of many economies of the region.

But in South Asia and East Asia, the contribution of expansion in arable land in raising regional water productivity will be almost negligible. The growth in water productivity will have to come from improved technical efficiency of water use, in terms of reduction in water depleted (consumed) per kilogram of the crop. This is in line with the projections by Cai and Rosegrant that water consumption per ha of land would reduce significantly for crops in South Asia and, while yield improves (South) East Asia (source: Cai & Rosegrant, 2003: figs. 10.9 and 10.10). This can also lead to water productivity improvements at the regional level. Another driver of water productivity improvement would be the improvements in genetic yield potential of the major food crops, such as wheat, maize, paddy, and sugarcane. The average wheat and paddy yield in East Asia has been increasing exponentially in the past two decades and is likely to continue (Bruinsma, 2009).

Both in India (South Asia) and China (East Asia), there is ample scope for improving irrigation, through interbasin water transfers (Kumar et al., 2012

---

[a] The scientific information about groundwater resource conditions is patchy at the local level, while some information about aquifer characteristics and recharge and abstraction are available at the regional level (Foster et al., 2006).

for India and Yang, 2003 for China) as both the countries are characterized by sharp regional variations in water resources endowment and degree of utilization of renewable water resources. India has most of its water resources in the GBM basin with limited demand for the same for agriculture in that basin (CWC, 2017). Its land resources are mostly in basins, which have very poor water endowment and where the stage of water development is already very high due to excessive demand for water in agriculture (Kumar et al., 2012). Similarly, in China, most of the water resources are in the South West region, which has very limited arable land. Whereas the water demands are very high in North China plains, which is acutely water-scarce but endowed with a vast amount of arable land. Nearly 40% of the agricultural production in China comes from this region (Yang, 2003). Hence, interbasin water transfer projects would be able to help expand irrigation growth in both South Asia and East Asia. We project it to be in the tune of at least 0.40% per annum in South Asia (instead of 0.1%) and the same figure in East Asia. Accordingly, pressure on water resources would be higher than what was projected by Alexandratos and Bruinsma (2012). At the country level, these interventions will impact agricultural water productivity by increasing the utilization of the water which otherwise goes into natural sinks for crop production.

Improvements in technical efficiency of irrigation water use would make major changes in water productivity both at the field level and basin/regional level in semiarid and arid water-scarce regions of South Asia and East Asia. Exceptions are areas where irrigation directly contributes significantly to groundwater recharge and base flows. Such areas exist in South Asia and East Asia, particularly in the canal commands.

The most ideal conditions for irrigation return flows contributing to groundwater recharge and baseflow are high rainfall, humidity, good hydraulic conductivity, and high water-table (Watt, 2008). The biggest impact is likely to come from India, and China, the two countries which are expected to invest mammoth sums in microirrigation systems (sprinklers and drips), improved water conveyance systems.

6.4 Climate change implications for future agricultural water productivity

The discussions in the previous section on future growth in agricultural outputs have not considered the effect of climate change. Given the global demand for agricultural commodities, the two important variables that drive the future land and water use in agriculture are the yield levels and water productivity attained in the crops. The earlier estimates of yield and output growth in different regions (Table 6.2) are based on a "business-as-usual" scenario. Drawing on the findings of the review presented in Chapter 4, the impact of climate change on crop yield and crop water productivity would depend largely on the nature of change in the regional or local climate such as a rise or lowering of temperature, changes in the maximum temperature, increase or decrease in the magnitude of precipitation, and extended or shortened duration of precipitation, the latitude of the region (mountains or midland or plains) and CO_2 concentrations (Kang et al., 2009). The types of challenges these climate-induced changes in crop yields and crop water productivity would pose to the sustainability of global agriculture and food security would depend a lot on how the regions/localities are positioned with respect to the availability of cultivable land and water resources.

For instance, if a region is land-scarce and water-abundant (like the eastern Gangetic Plains (EGP) in India and Bangladesh), a reduction in cereal yield due to climate change (say due to a rise in temperature) would matter a lot for regional food security, if the same is not offset by likely improvement in yield (of paddy in Bangladesh and wheat and paddy in the Indian part of EGP) due to CO_2 concentration. Basically, the problems of decline in grain production are not going to occur due to a decline in water productivity that might be witnessed due to yield reduction and an increase in ET. This is because the land is relatively scarce, and there is no scope for expanding the cultivated area to offset the yield reduction, whereas water is relatively abundant and the reduction in crop water productivity can be offset by an increase in the use of water.

On the contrary in the land-rich and water-scarce region of western and north-western India and the Indus basin area of Pakistan, if climate change impact is affected in the form of a rise in temperature (with no consequent increase in precipitation), then the impact on crop water productivity will be adverse and significant as the yield of the two major cereals, viz., wheat and paddy can reduce and their evapotranspiration rates can increase as a result of the anticipated changes in climate. This essentially means that more blue water (from deep aquifers) would eventually get used up for crop production. This can lead to the depletion of static groundwater in the region. However, in regions like peninsular India with hard rock aquifers, a reduction in precipitation will impact overall agricultural output significantly due to a reduction in annual availability of surface water (from reservoirs and diversion systems) and wells; and crop water productivity.

But if the same change is accompanied by an increase in precipitation, then the negative effect of temperature rise on yield might be subdued as shown by research (source: based on Akpalu, Hassan, & Ringler, 2008 and Popova & Kercheva, 2005). Further, there can be a positive effect in the form of yield gains for cereals viz., wheat and paddy, due to an increase in CO_2 concentration. On the other hand, the increase in rainfall during the monsoon season could, along with reducing the local temperature, increase the relative humidity thereby suppressing the ET. Further, higher rainfall can also increase soil moisture and renewable water availability (groundwater recharge and surface runoff). Under such a scenario, the impact on both crop water productivity and aggregate crop output will be positive and immense. This is because the available irrigation water from canal systems and the wells would be used for irrigating larger areas, as irrigation water requirement of crops would reduce due to an increase in effective moisture availability.

In the worst-case scenario of adverse climate impacts on crop water productivity through a negative effect on yield and positive effect on evapotranspiration, large parts of South Asia will be able to overcome them through technological interventions. Some of the technologies which can help substantially improve crop water productivity are microirrigation (MI) technologies, polyhouse, compost usage, drip irrigation (Conde, Ferrer, & Orozco, 2008), and mulching (Fuchs & Hadas, 2011; Xie, Wang, Jiang, & Wei, 2006; Yuan, Lei, Mao, Liu, & Wu, 2009). Khan, Hafeez, Rana, et al. (2008) explored the potential for water productivity improvements in the Liuyuankou Irrigation Area, China using groundwater flow modeling and concluded that the reduction in nonbeneficial evapotranspiration in irrigated cropland can make the extra water to be used in other areas. Conde et al. (2008) found, after analyzing rainfed maize with the CERES-Maize model in Mexico, that measures such as greenhouse construction, compost usage, and drip irrigation can be adopted to enhance adaptive capacities in maize growers.

Research is undertaken in different parts of Asia, including India, Pakistan, and China on the use of plastic mulch on a variety of food and cash crops, including rice in Pakistan (Jabran et al., 2015), maize in India (Ngangom et al., 2020) and China (Zhang, Wei, Han, Ren, & Jia, 2020; Zhang, Zhang, Qi, & Li, 2018) and vegetables such as tomato, capsicum and summer squash in Uttar Pradesh, India (Singh, Singh, & BHATT, 2014) show highly promising results in terms of increase in crop yield, reduction in water use and improvement in water productivity.

Climate change projections for Sub-Saharan Africa signals a warming trend, particularly in the inland subtropics; the frequent occurrence of extreme heat events; increasing aridity; and changes in rainfall—with a particularly pronounced decline in southern Africa and an increase in East Africa (Serdeczny et al., 2017). An increase in temperature and extreme heat events can lead to a potential reduction in yield of cereal crops such as maize if it is not accompanied by an increase in rainfall during the growing season that is capable of offsetting the adverse effects of temperature rise on cereal yields. However, this is an ideal situation where sufficient moisture is made available to the soils through supplementary irrigation in the event of dryness. This is unlikely to happen because a large proportion of the cultivated land in the region is rainfed lacking irrigation facilities, and because of this, atmospheric warming will lead to faster depletion of soil moisture and drought conditions. The result will be crop losses.

The serious implications of rainfall on SS Africa's agricultural output can be inferred from the empirical analysis by Barrios, Bertinelli, and Strobl (2010) which showed a positive correlation between SS Africa's annual GDP growth rates during the period from 1960 to 2000 and the rainfall anomalies, as the rainfall in the region mainly affects agricultural production and hydropower generation.

Cuculeanu, Tuinea, and Balteanu (2002) point out that soils of high water-holding capacity will be able to reduce the frequency of drought and improve the crop yield, in the event of declining water availability due to climate change. In that respect, SS Africa stands a poor chance for "cope up" due to degraded soils. With climate change, the growing period will reduce, and the planting date also needs to change for higher crop production. Climate change can decrease the crop rotation period, so farmers need to consider crop varieties, sowing dates, crop densities, and fertilization levels when planting crops (Cuculeanu et al., 2002) to adapt to climate-induced impacts on crop production. However, such adaptation measures are unlikely to emerge in Sub-Saharan Africa, especially in the least developed countries of that region, given the weak capacity for adaptation to climate change impacts in agriculture (Norford, 2009).

6.5 Findings and conclusions

Globally, demand for agricultural commodities is expected to grow due to population growth, and increasing per capita calorific requirements as a result of the increasing purchasing power of the people, in most regions of the world. The growth in demand is estimated to be 1.10% at the global level. However, there will be major variations in growth trends between regions due to the difference in population growth trends and the growth rate in per capita calorie intake. While population growth is likely to be zero or negative in many developed countries, the projected growth rate in average calorie intake is the lowest in these countries. The projected annual growth rate in calorie intake is highest for Sub-Saharan Africa, which had the lowest calorie intake in the base year (2005/07) followed by South Asia and East Asia. As a result, the growth rates in aggregate demand for agricultural outputs are high for these regions. The projected growth in demand for agricultural commodities is 2.40% for SS. Africa, 1.50% for North East/North Africa, and 1.70% for South Asia.

The agricultural outputs are also likely to increase globally at the rate of 1.10% to meet the growing demand for food and other commodities in the next three decades, with significant regional variations. The highest output growth is expected in Sub-Saharan Africa (2.40%), followed by South Asia (1.6%). This growth is largely consistent with the regional variation in demand growth. The factors that are expected to contribute to this growth will be different in different regions. In Sub-Saharan Africa, the expansion of arable land along with the creation of irrigation facilities will be a major contributor. In South Asia, the main contributor will be irrigation (0.40% per annum), and a lot of it is projected to be from interbasin water transfer projects such as the three gorges project in China and the Sardar Sarovar project in India. Increased agricultural production, which can be achieved mainly through the expansion of arable land that results in an effective increase in the utilization of green water from the natural catchments, will raise water productivity in agriculture in Sub-Saharan Africa. The increase in irrigation water use that would help increase the cropping intensities and crop yields can also result in water productivity improvements at both field-scale and basin-scale in South Asia and Sub-Saharan Africa. This is because when rainfed crops are irrigated, it leads to a disproportionate increase in yield and net income when compared to an increase in crop consumptive use of water. However, all these transitions would result in increased pressure on water resources, particularly in South Asia, North East, North Africa, and East Asia.

A rapid assessment of the potential impacts of climate change on crop production in south Asia shows varying impacts across regions. The land-rich areas of South Asia (in the Indo Gangetic plains in India and Pakistan) are largely equipped to deal with adverse trends in crop production-related climate variables such as temperature and rainfall with intensive use of available groundwater stocks, though such trends can reduce the water productivity of cereal crops. The hard rock regions of peninsular India can, however, witness large reductions in agricultural outputs in the event of yield drop and ET increase. But large parts of South Asia, especially the agriculturally prosperous areas, will be able to overcome the stress induced by climate through the use of precision irrigation technologies, viz., drips, polyhouse, and mulching for high-value fruits and vegetables. The land-scarce, water-rich regions of the eastern Gangetic Plains (in India and Bangladesh) will be adversely affected by any decline in cereal yields occurring in the events of temperature rise. At the same time, in the event of positive trends (of increased rainfall), the water-scarce and land-rich regions of south Asia will witness a substantial increase in production of cereals with the increase in the effective volume of water (blue water + green water) used, and improvements in crop water productivity resulting from increased yield and reduced ET.

Unlike South Asia, SS Africa will not be able to deal with the changes in temperature rise and extreme heat events on crop yields, if such events are not accompanied by an increase in rainfall, given the poor resilience of the agricultural production systems owing to inadequate irrigation infrastructure. Temperature rise can lead to faster moisture depletion and droughts, resulting in crop losses in the rainfed regions.

References

Akpalu, W., Hassan, R. M., & Ringler, C. (2008). Climate variability and maize yield in South Africa: Results from GME and MELE methods. In *Environment and production technology division IFPRI discussion paper* (pp. 1–12).

Alexandratos, N., & Bruinsma, J. (2012). *World agriculture towards 2030/2050: The 2012 revision.*

Barrios, S., Bertinelli, L., & Strobl, E. (2010). Trends in rainfall and economic growth in Africa: A neglected cause of the African growth tragedy. *Review of Economics and Statistics, 92*(2), 350–366.

Briceño-Garmendia, C., Smits, K., & Foster, V. (2008). Financing public infrastructure in sub-Saharan Africa: Patterns, issues, and options. In *AICD background paper 15, Africa infrastructure sector diagnostic.* Washington DC, USA: The World Bank.

Bruinsma, J. (2009). The resource outlook to 2050: By how much do land, water and crop yields need to increase by 2050. In *Vol. 2050. Expert meeting on how to feed the world* (pp. 24–26).

Cai, X., & Rosegrant, M. W. (2003). 10 World water productivity: Current situation and future options. In *Vol. 1. Water productivity in agriculture: Limits and opportunities for improvement* (pp. 163–178).

Central Water Commission. (2017). Reassessment of water availability in India using space inputs. In *Basin Planning and Management Organisation*. New Delhi: Central Water Commission.

Conde, C., Ferrer, R., & Orozco, S. (2008). Climate change and climate variability impacts on rainfed agricultural activities and possible adaptation measures: A Mexican case study. *Atmosfera, 19*(3), 181–194.

Cuculeanu, V., Tuinea, P., & Balteanu, D. (2002). Climate change impacts in Romania: Vulnerability and adaptation options. *Geology Journal, 57*, 203–209.

Food and Agriculture Organization of the United Nations. (2006a). *The state of food insecurity in the world 2006: Eradicating world hunger–taking stock ten years after the world food summit.* Rome: Food and Agriculture Organization of the United Nations.

Food and Agriculture Organization of the United Nations. (2006b). *World agriculture towards 2030/2050: Prospects for food, nutrition, agriculture, and major commodity groups.* Interim Report Rome: Food and Agriculture Organization of the United Nations.

Food and Agriculture Organization of the United Nations. (2007). Coping with water scarcity: Challenges of the 21st century. In *A World Water Day 2007 Brochure*. Rome, Italy: Food and Agriculture Organization.

Food and Agriculture Organization of the United Nations. (2009). *Report of the FAO expert meeting on how to feed the world in 2050.* Rome, Italy: Food and Agriculture Organization.

Food and Agriculture Organization of the United Nations. (2012). *Irrigation in eastern and southern Asia in figures: Aquastat survey 2011, FAO water report 37.* Italy, Rome: Food and Agriculture Organization.

Foster, V., & Briceño-Garmendia, C. (2010). 'Overview', in *Africa's infrastructure: A time for transformation*. Washington DC, USA: The World Bank.

Foster, S., Tuinhof, A., & Garduño, H. (2006). Groundwater development in Sub-Saharan Africa: Strategic overview of key issues and major needs. In *Case profile collection 15*. Washington, DC: The World Bank.

Fuchs, M., & Hadas, A. (2011). Mulch resistance to water vapor transport. *Agricultural Water Management, 98*(6), 990–998.

Goyal, A., & Nash, J. (2016). *Reaping richer returns: Public spending priorities for African agriculture productivity growth.* African Development Forum, World Bank Group.

Jabran, K., Ullah, E., Hussain, M., Farooq, M., Zaman, U., Yaseen, M., & Chauhan, B. S. (2015). Mulching improves water productivity, yield and quality of fine rice under water-saving rice production systems. *Journal of Agronomy and Crop Science, 201*(5), 389–400.

Kang, Y., Khan, S., & Ma, Y. (2009). Climate change impacts on crop yield, crop water productivity and food security—A review. *Progress in Natural Science, 19*(2009), 1665–1674.

Khan, S., Hafeez, M. M., Rana, T., et al. (2008). Enhancing water productivity at the irrigation system level: A geospatial hydrology application in the Yellow River Basin. *Journal of Arid Environment, 72*, 1046–1063.

Kumar, M. D. (2007). *Ground water management in India' physical, institutional and policy alternatives.* New Delhi: Sage Publications.

Kumar, M. D. (2014). *Changes in water productivity in agriculture with particular reference to South Asia: Past and the future.* New Delhi: Final report submitted to International Water Management Institute.

Kumar, M. D., Bassi, N., & Singh, O. P. (2020). Rethinking on the methodology for assessing global water and food challenges. *International Journal of Water Resources Development, 36*(2–3), 547–564. https://doi.org/10.1080/07900627.2019.1707071.

Kumar, M. D., Sivamohan, M. V. K., & Narayanamoorthy, A. (2012). The food security challenge of the food-land-water nexus in India. *Food Security, 4*(4), 539–556.

Namara, R. E., Gebregziabher, G., Giordano, M., & De Fraiture, C. (2013). Small pumps and poor farmers in Sub-Saharan Africa: An assessment of current extent of use and poverty outreach. *Water International, 38*(6), 827–839.

Ngangom, B., Das, A., Lal, R., Idapuganti, R. G., Layek, J., Basavaraj, S., ... Ghosh, P. K. (2020). Double mulching improves soil properties and productivity of maize-based cropping system in eastern Indian Himalayas. *International Soil and Water Conservation Research, 8*(3), 308–320.

Norford, E. (2009). Adaptation to climate change in Sub-Saharan Africa: An investigation of capacity-building and national adaptation programs of action. Honours Thesis In *Environmental studies* Connecticut College.

Popova, Z., & Kercheva, M. (2005). CERES model application for increasing preparedness to climate variability in agricultural planning-risk analyses. *Physics and Chemistry of the Earth, 30*, 117–124.

Qureshi, A. S., McCornick, P. G., Sarwar, A., & Sharma, B. R. (2010). Challenges and prospects of sustainable groundwater management in the Indus Basin, Pakistan. *Water Resources Management, 24*(8), 1551–1569.

Rockström, J., Barron, J., & Fox, P. (2002). Rainwater management for increased productivity among small-holder farmers in drought prone environments. *Physics and Chemistry of the Earth, Parts A/B/C, 27*(11–22), 949–959.

Rosa, L., Chiarelli, D. D., Rulli, M. C., Angelo, I. D., & D'Odorico, P. (2020). Global agricultural economic water scarcity. *Science Advances, 6*, eaaz6031.

Serdeczny, O., Adams, S., Baarsch, F., Coumou, D., Robinson, A., Hare, W., ... Reinhardt, J. (2017). Climate change impacts in Sub-Saharan Africa: From physical changes to their social repercussions. *Regional Environmental Change, 17*, 1585–1600.

Singh, V. P., Singh, P. K., & BHATT, L. (2014). Use of plastic mulch for enhancing water productivity of off-season vegetables in terraced land in Chamoli district of Uttarakhand, India. *Land Capability Classification and Land Resources Planning Using Remote Sensing and GIS 3, 13*(1), 68–72.

Watt, J. (2008). *The effect of irrigation on surface-ground water interactions: Quantifying time dependent spatial dynamics in irrigation systems.* Doctoral dissertation Charles Sturt University.

Xie, Z., Wang, Y., Jiang, W., & Wei, X. (2006). Evaporation and evapotranspiration in a watermelon field mulched with gravel of different sizes in Northwest China. *Agricultural Water Management, 81*(1–2), 173–184.

Yang, H. (2003). Water, environment and food security: A case study of the Haihe River basin in China. *WIT Transactions on Ecology and the Environment, 60.*

You, L., Wood, S., & Sichra, U. W. (2007). *Generating plausible crop distribution and performance maps for Sub-Saharan Africa using a spatially disaggregated data fusion and optimization approach.* Discussion Paper 00725 Energy and Production Technology Division, International Food Policy Research Institute.

Yuan, C., Lei, T., Mao, L., Liu, H., & Wu, Y. (2009). Soil surface evaporation processes under mulches of different sized gravel. *Catena, 78*(2), 117–121.

Zhang, P., Wei, T., Han, Q., Ren, X., & Jia, Z. (2020). Effects of different film mulching methods on soil water productivity and maize yield in a semiarid area of China. *Agricultural Water Management, 241*, 106382.

Zhang, F., Zhang, W., Qi, J., & Li, F. M. (2018). A regional evaluation of plastic film mulching for improving crop yields on the Loess Plateau of China. *Agricultural and Forest Meteorology, 248*, 458–468.

CHAPTER 7

Past growth in agricultural productivity in South Asia

7.1 Introduction

Asian agriculture had undergone an enormous transformation during the past 6 decades beginning with the Green revolution, marked by a quantum jump in the yield of cereal crops. Though South Asia lagged behind South East Asia and East Asia, due to structural hurdles, it overcame these and picked up accelerated growth later (Vos, 2018). The agricultural output had increased remarkably in the last 40 years (Alexandratos & Bruinsma, 2012). Yield growth for major food crops viz., paddy and wheat has been quite significant for most countries, viz., India, Pakistan, Bangladesh, and Nepal (Timsiha & Connor, 2001), though it has been low in Afghanistan which remains a food-insecure country heavily dependent on food aid due to civil war and natural disasters. Yield growth of major cereals resulting from genetic improvement, increase in cropping intensity, enabled by irrigation expansion (Pingali, 2012), and to a very limited extent, expansion in arable land triggered agricultural growth in South Asia.

The Green revolution, which introduced high-yielding varieties of major cereals such as wheat and rice in South Asian countries along with chemical fertilizers, was also marked by irrigation expansion (Pingali, 2012), especially that of well irrigation. The most dramatic increase in well irrigation occurred in India, which was propelled by rural electrification, easy access to pumps and drilling technologies, institutional financing for digging/drilling wells, and supported by enhanced knowledge about aquifers. Heavily subsidized electricity for agricultural groundwater pumping also made pumping water from large depths economically feasible for the farmers. An explosion in well irrigation technology-enabled increased access of millions of farmers in the semiarid and arid regions, who were otherwise largely deprived of reliable irrigation given the limited reach of public

Current Directions in Water Scarcity Research, Volume 3
https://doi.org/10.1016/B978-0-323-91277-8.00012-5
137

irrigation in these regions, to small scale irrigation with a reasonable degree of dependability depending on the geohydrological settings. Over the years, well irrigation also expanded in Pakistan (Qureshi, McCornick, Sarwar, & Sharma, 2010) and Bangladesh (Qureshi, Ahmed, & Krupnik, 2014).[a]

Canal irrigation encouraged farmers to cultivate highly water-intensive crops such as sugarcane, paddy, and wheat. But farmers with access to well irrigation went for irrigated dry crops and cash crops, such as maize, cotton, castor, cumin, groundnut, pearl millet, jowar, potato, onion, and other vegetables. In Andhra Pradesh, which has one of the largest networks of gravity irrigations, 75% of the gravity-irrigated area was under paddy, and 92% of the paddy irrigation was covered by gravity irrigation. In some of the largest irrigation commands in Indian and Pakistan Punjab, well irrigation supplemented canal irrigation for the paddy-wheat system. Extensive well irrigation infrastructure brought some of the most unproductive lands in the region under cultivation, thereby resulting in the expansion of cultivated land.[b]

Apparently, the amount of crop output per unit of canal water supplied from large surface irrigation systems that took water to distant farms was much lower than that of wells that provided irrigation to the nearest fields. However, canal water augmented recharge to shallow aquifers in many areas through seepage and irrigation return flows (especially in irrigated paddy), and the same was recycled back by wells. Canal water supplied in large quantities ensured the sustainability of well irrigation to a great extent (Jagadeesan & Kumar, 2015; Kumar & Perry, 2019). Such irrigation water recycling systems exist in Indian Punjab, IBIS in Pakistan, and parts of western and peninsular India. In many semiarid and arid areas with low to medium rainfall, where farmers depended entirely on well irrigation, overexploitation of the resource led to depletion. In any case, all these led to the large-scale expansion of irrigation in South Asia, with intensive cropping in many areas, with the exception of Afghanistan. The total net irrigated area in the subcontinent (excluding Afghanistan and Iran) increased from 44.91 m

[a] In Pakistan, it made inroads into the command areas of Indus Basin Irrigation System, with farmers using it as a source to supplement surface water supplied through the gravity irrigation (Qureshi et al., 2010). Bangladesh witnessed a dramatic increase in the number of shallow and deep tube wells during the period from 1982 to 2013 (Qureshi et al., 2014).
[b] An example is the Thar desert in western Rajasthan, which now has large areas with desert soils with very low productivity under well irrigation. Several irrigated dry crops and oil seed crops such as pearl millet, sorghum, cow pea, castor, cumin and fennel, cluster bean are cultivated.

per ha in 1962 to 81.83 m per ha in 1998, with the highest growth witnessed in India, Sri Lanka, and Pakistan (Barker & Molle, 2004). Most river basins in the semiarid and arid belt regions (Indus, Cauvery, Krishna, Sabarmati, west-flowing rivers of India north of Sabarmati, Pennar, etc.) were over-appropriated. Groundwater intensive use in upper catchments of some river basins of India was often at the cost of reduced stream flows (Kumar, 2010), which even impacted the surface systems such as irrigation tanks and reservoirs (Kumar & Vedantam, 2016; Talati & Kumar, 2019).

The creation of irrigation infrastructure, both private and public, continue to encourage farmers who were earlier practicing low input agriculture in the region, to intensify the use of external inputs and use of high yielding varieties and also shift to high-value cash crops. Such a trend was more visible in the central Indian states, which constitute India's tribal heartland. An overall increase in agricultural outputs, combined with crop diversification toward high-value commodities such as fruits, vegetables, milk and fish, driven by growing demand resulting from rising per capita income, and development of market infrastructure, urbanization, and technological improvements resulted in growth in value of agricultural outputs in the region (Joshi, Gulati, Birthal, & Tewari, 2004). Such a trend was quite significant in India during the past one and a half decades (Rada, 2013). In this chapter, we will systematically analyze the determinants of past growth in agricultural outputs of South Asia, and their implications for agricultural water productivity.

7.2 Past trends in agricultural water productivity in South Asia

South Asian region is characterized by extreme spatial variability in climate and agro-ecologies—from hot and hyperarid (Thar desert of Rajasthan) to hot and arid to hot and semiarid to hot and humid to cold and humid (Northeastern India, Bhutan). South Asia is also known for the highest cropping intensities among all regions in the world (Siebert, Portmann, & Döll, 2010).

The subcontinent is endowed with regions of water abundance and regions of water scarcity. But, the unique characteristic of the subcontinent is that the regions which have water in plenty have very limited land resources, like the deltaic plains of Bangladesh and Eastern Gangetic plans in India. On the contrary, the regions which have plenty of arable lands, have limited water resources, which put major constraints in their ability to raise agricultural production, like North western India, the Indus basin area of Pakistan, most parts of Afghanistan, and most parts of South Indian Peninsula. Such patterns are sharp in India (Kumar, Sivamohan, & Narayanamoorthy, 2012).

7.2.1 Population growth and change in consumption pattern

While the population of South Asia has more than doubled from 708 million in 1970 to 1520 million in 2006, with per capita income increasing remarkably, the demand for food increased modestly in terms of consumption (kilogram) per capita per annum. Notably, the most remarkable increase in consumption has come from milk and milk products, with the average per capita consumption increasing 1.8% annual growth. Whereas in the case of cereals, the change hasn't been consistent, with per capita consumption declining, increasing and then declining in 2005–2007 to a level somewhere near the 1969–1971 levels. The sharp decline in 1979–1981, despite no increase in consumption of most other food items and some decline in consumption of pulses, which is not in conformity with the trend observed in (all) developing countries during that period (see table 2.4 of Alexandratos & Bruinsma, 2012), could be attributed to the increase in regional food prices, resulting from a shortage.

From Table 7.1, it is also evident that there has been a consistent growth (1.85% per annum) in vegetable oils also, and a 1% growth in consumption of other food items. In calorific terms also, the increase is modest (from 2072 to 2293 kcal/capita/day), mainly due to the differences in the consumption levels of meat in developed countries.[c]

7.2.2 Change in agricultural outputs

The agricultural outputs have recorded a quantum jump in the subcontinent during the 36-year period, through expansion in arable land (net sown area), crop intensification (double and triple cropping), productivity (yield) improvement, and improvements in harvesting techniques, to meet the growth in demand resulting mainly from population growth and modest growth in per capita calorie consumption. While arable land expansion contributed only 6% in crop production during 1961–2007, yield improvement contributed 82% during the same period and cropping intensity

[c] As noted by Alexandratos and Bruinsma (2012), most developing countries including India are not following a rapid transition pattern in commodity composition of food consumption. While for some of them, it is a question of slow gains in income and persistent poverty, for some others, it is a question of nonchanging food habits.

Table 7.1 Change in consumption of food in South Asia (1969–1971 to 2005–2007) in kilogram/annum.

Type of food	Per capita consumption				Annual compounded growth rate (%) during 1969–1971 to 2005–2007[a]
	1969–1971	1979–1981	1989–1991	2005–2007	
Cereals	151.0	147.0	161.00	152.00	0.00
Roots and tubers	17.0	20.00	18.00	25.00	1.08
Sugar and sugar crops	20.70	19.40	20.60	19.10	−0.22
Dry pulses	14.50	11.10	12.00	10.30	−0.94
Vegetable oils, oil seeds and products	4.60	5.70	7.10	8.90	1.85
Meat	3.90	4.00	4.60	4.40	0.33
Milk and dairy (excluding butter)	38.0	42.00	54.00	71.00	1.75
Other food (kcal/person/day)	90.0	87.00	98.00	137.00	1.10

[a]Authors' own estimates based on the data provided in the previous columns.
Source: Alexandratos and Bruinsma (2012).

contributed 12% (Alexandratos & Bruinsma, 2012, table 4.12). This basically meant that agricultural output growth came with additional pressure on water resources, as not only cropping intensity increased but yield improvement through high yielding varieties also required more water (Kumar, Viswanathan, & Bassi, 2014).

While the demand for agricultural commodities (all commodities and all uses) increased by 3% per annum in South Asia in the period from 1970 to 2007, the production also increased by 3%. Sector-wise break up shows that in the case of cereals, there is a trade balance of around 2.7 million ton; in the case of meat, the balance is 0.5 m per ton; and in the case of milk, the balance is −0.61 million liters of liquid milk equivalent, indicating an imbalance. This is in spite of impressive growth of 4.4% per annum in milk production in the region. Within agriculture, crop production (which includes oilseeds, sugar, roots, and tubers) recorded a 2.6% annual growth during the 46-year period from 1961–2007 (source: based on table 4.3 in Alexandratos & Bruinsma (2012)).

Crop production is the mirror of agricultural outputs. Higher production of livestock-based products such as chicken, beef, milk, and other milk products requires more land for the cultivation of fodder (for animals), and cereals (for poultry and animals). The livestock-based products would also consume more water per calorie of energy produced than crop-based products such as cereals and vegetables (Mekonnen & Hoekstra, 2012).

7.2.3 Changes in yield, cropped area, and irrigated area

Among the three factors responsible for raising crop production, expansion in arable land (net sown area) and cropping intensities happened more in the naturally water-scarce, land rich regions, and less in the land-scarce, water-rich regions. Yield growth has occurred both in water-scarce and water-rich regions, though at a higher rate in the former. However, many regions, which have a good amount of arable land, continue to record poor yields for major crops due to inadequate irrigation facilities. In other regions, irrigation development, particularly the development of well irrigation had substantially contributed to increased cropping intensity and a minor increase in arable land under cultivation. Yield enhancement has also occurred simultaneously with the replacement of short-duration rain-fed varieties with irrigated high-yielding varieties which are of long duration—like in paddy, wheat, and maize.

Two important factors had acted as major drivers of agricultural growth. The first one is that a major share of the gross cropped area of the subcontinent is under cereals, viz., wheat, maize, and paddy, whose contribution to cropping, both in percentage and aggregate terms, had increased significantly, and the second one was the large scale introduction of high yielding varieties of these crops during the past 40 years under the Green revolution, supported by increased irrigation facilities and extensive and intensive use of chemical fertilizers. As a result, yield of these crops increased remarkably. For example, wheat yields in South Asia increased by an average of 54 kg per annum over 1980–2012. Maize yields increased by 48 kg per annum; paddy yield by 44 kg; and sugarcane yield by 590 kg (source: author's own estimates based on FAOSTAT).

The current average yield and historical growth in yield of major crops, viz., maize, paddy, sugarcane, and wheat for the countries of South Asia are given in Table 7.2. High yield growth was experienced for wheat in Bhutan, India, Iran, and Nepal, and moderate growth in Pakistan and Afghanistan. High yield growths in paddy were noticed in Bhutan, India, Iran, and Nepal. The growth, witnessed in paddy yields in Bangladesh and Pakistan, is moderate. Currently, average wheat yields are highest in India, Pakistan and Bangladesh. Among these countries, Bangladesh has a relatively small area under wheat. As regards paddy, the average yields are highest in Iran, followed by Sri Lanka, India, Bhutan, and Pakistan. Bangladesh and Afghanistan have low average yields in Paddy. Sugarcane yield is highest in Iran, followed by India. For other countries, the yield levels are far below the Iranian average. Maize yield is highest in Bangladesh, followed by Pakistan.

The following factors contributed to yield growth for the major crops of South Asia: gradual replacement of rain-fed varieties by long duration and high yielding irrigated varieties; supplementary irrigation of crops which were earlier grown under rain-fed condition, leading to realizing of the full farm yield potential; introduction of crop varieties, with higher genetic yield potential for wheat (Pingali, 1999; Siddique, Loss, Herwig, & Wilson, 1996) and rice (Tran, 1997); improvement in harvest index (Siddique, Tennant, Perry, & Belford, 1990; Tran, 1997). Needless to say, in different regions, the contribution of these factors to yield growth has been differential. In regions, which already had a high level of irrigation (like Haryana and Punjab in India and IBIS command areas of Pakistan), most of the yield growth came from genetic improvements in yields, particularly for wheat, paddy, and maize. In other regions, which had low levels of irrigation (peninsular India, Bangladesh, and Western India), expansion in irrigation, particularly well irrigation, enabled supplementary irrigation of crops grown

Table 7.2 Current yields and yield growth for major crops in South Asian countries.

Country name	Current yield of major crops (kg/ha)				Annual growth (%) in yield of crops during 1980–2012			
	Maize	Paddy	Sugarcane	Wheat	Maize	Paddy	Sugarcane	Wheat
Afghanistan	2199	2439	25,000	2010	0.86	0.36	0.87	1.48
Bangladesh	6584	2933	42,771	2779	7.19	1.17	−0.1	1.2
Bhutan	2300	3480	31,111	2143	1.57	1.75	0.31	2.4
India	2507	3591	68,344	3173	2.44	1.85	1.02	2.51
Iran	3494	5000	85,714	1971	2.83	1.79	0.69	2.2
Nepal	2501	3312	45,447	2412	1.36	1.7	3.09	2.21
Pakistan	4268	3482	55,829	2709	3.88	1.14	1.19	1.72
Sri Lanka	3419	3885	53,408		3.63	1.28	0.69	NA

Source: Author's estimates based on FAOSTAT for the period from 1980 to 2012.

under rain-fed conditions. Among these factors, the last three factors contribute to water productivity at the field level.

But the naturally water-scarce regions, which are well endowed in arable land, have driven the agricultural growth, rather than the water- rich regions. The reason is that there are serious limits for the latter to enhance crop outputs due to agroecological factors such as floods, waterlogging and salinity, short winter (Ladha, Fischer, Hossain, Hobbs, & Hardy, 2000), and land scarcity (Kumar et al., 2012), which create poor environmental conditions for crop growth. Due to high rainfall and humidity, many crops are grown under rain-fed conditions, wherein it is difficult to achieve good farm management, with proper agronomic inputs and water control due to excessively high rainfalls (West Bengal, Assam, Eastern UP, coastal Orissa, and Bihar in India, and Bangladesh). The sharp differences in yield realized in wheat and paddy and also the yield gap between Punjab in northwestern Indian and eastern Gangetic plains in west Bengal illustrate this point (Pathak et al., 2003).

While water is a major limiting factor for raising agricultural production in the land-rich regions receiving low rainfall, the tropical climate in such regions provides the best environmental conditions for achieving high yields. Irrigation expansion—from 37 m per ha to 90 m ha from 1961/1963 to 2005/2007—enabled this growth (Alexandratos & Bruinsma, 2012). As a matter of fact, part of the irrigation expansion in the subcontinent took place on arid and hyperarid land which is not suitable for rain-fed agriculture. Indus basin area in Pakistan, north western India (Punjab and Haryana), and Western Rajasthan are examples. In some regions and countries of South Asia, irrigated arid and hyperarid landforms 16 out of 90 million ha of the irrigated land presently in use in South Asia. Most of the remaining irrigated lands in these countries fall in semiarid regions (Alexandratos & Bruinsma, 2012).

South Asia still has around 55 m per ha of uncultivated land, which is suitable for cropping. The rate of increase in the use of arable land in the region has been very low (0.14% per annum) during 1961/1963– 2005/2007. The total arable land in use for crop production is 140 m per ha, and the net area available for future expansion is only 11 m per ha, with nearly 43 m per ha under forests, protected areas, and built-up areas (Fischer, Hizsnyik, Prieler, & Wiberg, 2011). The land areas, which remain uncultivated, are those where rainfall is very low. There is almost no scope for expanding the net sown area unless an irrigation facility is created.

Fig. 7.1 shows the trend in net cultivated area and net irrigated area in South Asia. The net irrigated area in South Asia expanded from 76.19 m

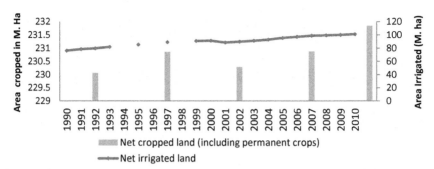

Fig. 7.1 Trends in cropped area and irrigated area in South Asia (1991–2011).

per ha to 101.10 m per ha during 1990–2010, while the total cropped area (net sown area) increased from 230 m per ha to 231.8 m per ha, in nearly 2 decades, indicating almost nil growth. The highest expansion in the net irrigated area was witnessed in Bangladesh (122% growth), followed by India (46%), Nepal (29%), and Pakistan (27%) (source: author's own estimates based on FAOSTAT for the years from 1991 to 2011).

7.2.4 Changes in agricultural water productivity

The factors that can drive water productivity at the farm level are changes in the composition of the farming system from water-intensive, low-value crops to low water-consuming, high-value crops (Kumar & van Dam, 2013); improved allocation of irrigation water to crops to meet the gap between PET and effective rainfall; use of efficient irrigation technologies, though in selected localities (Kumar & van Dam, 2013); controlled irrigation or deficit in situations where the excessive application of water to crops (irrigation dosage corresponding to the descending side of the yield response curve) is resulting in low yields (Deng, Shan, Zhang, & Turner, 2006; FAO, 2003; Zhang, 2003); adoption of crop varieties which are fast maturing or have lower leaf area index or are more efficient in capturing solar radiation that reduces evaporation/ET ratio (Siddique et al., 1990); or crops having high "transpiration efficiency" (kg/T).

Among these, the following factors had facilitated water productivity improvements in South Asian agriculture, at the farm level. First, large-scale well development and expansion in groundwater irrigation in the entire South Asia region, particularly in India, Pakistan, and Bangladesh, had enabled supplementary irrigation of Kharif crops, which largely depend on erratic monsoon rains, and irrigated winter crops such as wheat and maize to realize the optimum yield. The development of wells in command areas had also helped improve the adequacy and timeliness of water delivery to crops in such areas, which otherwise has low reliability. Vast areas served by the Indus Basin Irrigation System in Pakistan (13.0 m per ha) and large command areas in Punjab and Haryana in India are benefited by well irrigation (Qureshi et al., 2010), which supplements gravity irrigation from canals. The farmers who depend on both wells and canals in these regions obtain much higher crop yields and income (GOP, 2012), and therefore water productivity is likely to be high. On the other hand, well irrigation has become widespread in India, Bangladesh, and Pakistan, improving water productivity in crop production in many ways. First, it enabled farmers to grow irrigated cash crops, replacing rain-fed cereal crops. It also enabled them to provide timely irrigation to crops in a controlled fashion. While timeliness and adequacy improved yields, water control helps improve the consumptive use fraction of applied water and ET fraction of the total consumptive use (Kumar & van Dam, 2013).

The next alternative to improve water productivity at the farm level is through allocative efficiency improvements. In regions, where the farmers are confronted with a high marginal cost of using water, they are already allocating irrigation water to high-value crops, which generates higher income per unit volume of water. Examples are the millions of diesel pump owners and water buyers (of electric and diesel pumps) in the Indo-Gangetic plains (Kumar, Singh, & Sivamohan, 2010). The well irrigators in the IBIS, who are using diesel pumps,[d] are also confronted with the high marginal cost of pumping water (Qureshi et al., 2010), and therefore are also expected to be achieving high farm-level water use efficiency by allocating more water to crops which give higher return per unit volume of water. However, low allocative efficiency would be prevalent in areas of South Asia where farmers are fully dependent on cheap canal water, most importantly in the IBIS and canal command areas in India. Low allocative efficiency is also expected among farmers who practice well irrigation, using subsidized electricity supplied by the State Electricity Boards. By the most conservative estimate, this could be anywhere near 25–30 m per ha in South Asia.

The key strategies for improving water productivity at the basin/regional level, regardless of whether the crop is grown under rain-fed or irrigated conditions, are (i) increasing the marketable yield of the crop; (ii) reducing all basin outflows (drainage outflows, seepage, and percolation into saline formations) and evaporative outflows other than the crop stomatal

---

[d] In fact, they constitute a major share of the well irrigators in the IBIS (Qureshi et al., 2010).

transpiration; and (iii) increasing the effective use of rainfall, stored water, and marginal quality water (FAO, 2003; Molden, Murray-Rust, Sakthiva-divel, & Makin, 2003). Interregional water allocation, with the relative expansion of cropland in regions which receive high-intensity solar radiation, low aridity (Abdullaev & Molden, 2004; Zwart & Bastiaanssen, 2004), and nutrient-rich and salt-free soils (Schmidhalter & Oertli, 1991) are other important strategies.

Among these, the following changes had driven agricultural water productivity growth in South Asia. First, the amount of water utilized for irrigation from many river basins of South Asia, particularly, Indus, Ganges, Helmand, and many river basins of western India, south Indian peninsula and central India had increased substantially, through (large, medium, and small) storage and diversion systems. Water development stress is very high in most of these basins, except Ganges (Kumar et al., 2014). There are very few basins in South Asia, which have significant drainage outflows into the ocean. They include Godavari, Brahmani, Ganges, Brahmaputra and Meghna (Kumar, Patel, Ravindranath, & Singh, 2008). Of these, the last three rivers, which constitute the GBM system, have their drainage outflows from Bangladesh. But, the potential for utilization of this untapped water will be quite low, unless water is transferred to water-scarce, land rich basins for increasing the utilization through irrigation expansion.[e] In Pakistan, the drainage outflow from Indus is almost absent, and the renewable water resources of the basin are almost fully tapped.

Second: as regards measures to reduce percolation and evaporation losses, they would be useful only in situations where the deep percolation water is nonrecoverable (Allen, Willardson, & Frederiksen, 1997). In canal command areas such as the IBIS in Pakistan (FAO, 2003; GOP, 2012; Qureshi et al., 2010), and Bhakra command in Punjab (FAO, 2003), Mahi command in South Gujarat and Mulla Command in Maharashtra of India, there is extensive recycling of the seepage and percolation (Kumar, Turral, Sharma, Amarasinghe, & Singh, 2008). In Sri Lanka, An IWMI study of an irrigation system in Kirindi Oya in southern Sri Lanka found that at the system level crops consumed only 23% of the total water supply from the system, including both rainfall and external irrigation water. Of the remainder, 8% was

[e] Barring one or two basins, there isn't much arable land available in these basins (especially GBM) which can be put to cultivation using the water resources. The only option is to increase the cropping intensity, possibility of which is again low due to high rainfalls, water-logging and floods.

used for grazing land, 6% evaporated from the reservoir, 16% was lost to the sea, 3% drained into lagoons, while as much as 44% of the water supply went to perennial vegetation that had developed since the starting of the scheme, from irrigation seepage and recharge of the shallow groundwater.

But there are regions and situations in South Asia where the seepage and percolation from irrigated land will not be fully available as return flows to groundwater, and where bare soil evaporation, which does not contribute to plant growth, will be significant.[f] Reducing these losses can lead to real water saving at the basin level (Kumar & van Dam, 2013). Geographical areas with such physical conditions are large. Under the conventional method of irrigation, theoretically, these losses could be significant in such areas. But, in most regions receiving extensive flow irrigation, water tables are generally high. Therefore, our focus should be on arid and semiarid regions that receive groundwater irrigation. Such regions include western India (semi-arid and arid parts of Gujarat, Rajasthan, MP, and Maharashtra), peninsular India (excluding Kerala), dry-land areas of Sri Lanka (which receive relatively low rainfall), Afghanistan, and parts of arid Pakistan which are not under IBIS. The irrigation technologies which can reduce these losses and improve water productivity are drips, sprinklers, and mulching.

Adoption of these irrigation technologies is poor in South Asia. As per the statistics compiled by ICID, among all the eight South Asian countries, India is the only country that has crops under sprinkler and drip irrigation, accounting for 8.1% of the net irrigated area. The total area under drips and sprinklers in the country is 4.94 m per ha. Of this, only 1.89 m per ha is under drips (see Table 7.3).

Substantial areas under high value crops such as cotton, castor, groundnut, potato, sugarcane, banana, grapes, vegetables, water melon, mango, lemon, and pomegranate, crops to which drip technology is most amenable, are already irrigated by drips (Kumar, Turral, et al., 2008). As pointed out by Kumar, Turral, et al. (2008), the ultimate potential for water-saving MI systems is very low in India (5.89 m per ha),[g] as against the projections by the Planning Commission (of 97 m per ha), due to a wide variety of physical, socioeconomic and institutional and policy constraints, preventing the

---

[f] The former situation is one in which the aridity is high and groundwater table is deep (Kumar & van Dam, 2013), whereas the latter occurs when the aridity is high and the entire wet soil is not covered by canopy and part of it exposed to solar radiation flux for long time periods (Siddique et al., 1990).

[g] Here, the area under field crops such as wheat, pearl millet, sorghum, mustard, etc., which can be irrigated by sprinklers are not considered, as such uses may not result in water saving.

Table 7.3 State-wise area under drip and sprinkler irrigation systems in India (2016–2017).

Name of state	Area under (in ha)		Total area (ha)
	Drip	Sprinkler	
Andhra Pradesh	851,000	256,000	1,107,000
Maharashtra	659,000	271,000	930,000
Gujarat	457,000	471,000	928,000
Karnataka	397,000	519,000	916,000
Rajasthan	146,000	680,000	826,000
Madhya Pradesh	254,000	112,000	366,000
Tamil Nadu	224,000	23,000	247,000
Chhattisgarh	15,000	159,000	174,000
Odisha	23,000	86,000	109,000
Bihar	8500	94,000	102,500
Haryana	28,000	48,000	76,000
Uttar Pradesh	15,000	47,000	62,000
Punjab	30,000	3000	33,000
Jharkhand	16,000	14,000	30,000
Kerala	12,000	8000	20,000
Uttarakhand	3622	2059	5681
Himachal Pradesh	2536	3116	5652
Mizoram	2242	1364	3606
Sikkim	1558	1626	3184
Goa	440	547	987
West Bengal	439	459	898
Tripura	444	280	724
Meghalaya	308	307	615
Nagaland	444	0	444
India total	*3,148,038*	*2,800,815*	*5,948,853*

Source: https://pmksy.gov.in/microirrigation/AtGlance.aspx.

adoption of this technology. Hence, the future potential for water-saving MI systems such as drips remains quite low in India. In Pakistan, the dominance of the rice-wheat system prohibits the adoption of drip systems in IBIS, even if the farmers want to adopt them for yield gains, due to poor economic viability. In Bangladesh, a large proportion of the irrigated crop land is under paddy, and only the small vegetable growers can be the potential adopters of MI systems.

Third, the regions which have a high climatic yield potential of rice and wheat such as Indian Punjab and Haryana (Pathak et al., 2003) and Indus basin irrigation command in Pakistan, where the actual yield realized in the field is very high, and water productivity (kg/ET) is high (Zwart & Bastiaanssen, 2004), there is intensive cultivation of these two major cereals. Third, the arable land area has also expanded with them being utilized for producing monsoon crops, and only 11 m per ha of arable land is left unutilized (Fischer, van Velthuizen, & Nachtergaele, 2009). This indicates the effective utilization of rainwater. There is also extensive utilization of saline groundwater for crop production through blending with surface water from canals with pumped groundwater from the saline formations underlying the basin, particularly in the IBIS (Qureshi et al., 2010).

As regards effective use of rainfall in the region, there hasn't been much expansion in the arable land in South Asia, and the region has the highest cropping intensity—the ratio of the amount of land under crops and the total area available for cultivation—in the world. Hence, increase in the use of water in the soil profile for crop production at the regional level hasn't been significant enough to contribute to regional water productivity improvements.

The growth trajectory in the value of (all) agricultural outputs from seven South Asian countries for the period from 1980 to 2012 is given in Fig. 7.2. The growth rate was 3.15%. But the growth is not uniform across the subcontinent. The highest growth was recorded in Pakistan (3.5%), while the lowest growth is in Sri Lanka (1.026%), followed by Afghanistan (1.27%). India recorded an impressive growth rate of 3.14%, which is almost the same as the South Asian average (source: Author's own estimates based on FAOSTAT).

The highest growth rate in Pakistan could be attributed to sustained expansion in irrigated areas and an increasing shift toward high-value crops

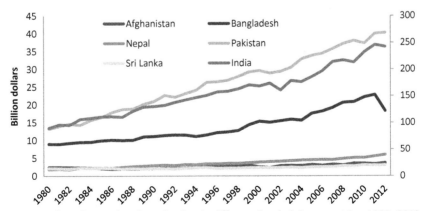

Fig. 7.2 Value of agricultural production in different South Asian countries: 1980–2012.

such as cotton and sugarcane. This was enabled by tapping of the naturally available groundwater and the recharge from canal seepage and irrigation return flow, through extensive development of tube wells in the command areas of the Indus Basin Irrigation System. The Indus Basin Irrigation System has become the most efficient water recycling machine. The livestock production system, which was supported by by-products of wheat and paddy grown on a large scale in the country helped enhance the value of agricultural outputs. In India, increased cropping intensity through irrigated area expansion driven by well developed, crop diversification with increased area under high-value crops such as oilseeds, fibers, fruits and vegetables, and composite crop-dairy farming and rise in the price of agricultural commodities in real terms had boosted growth in agricultural outputs in value terms.

If we assume that at the South Asia level, the extent of mining of groundwater resources annually is not significant in comparison to the total annual groundwater draft,[h] the implications of agricultural production growth for regional water productivity could be as follows. The water productivity in agriculture production has increased at a rate of more than 3% annually in South Asia during the past 30 years, with the highest growth in Pakistan, followed by India. While the annual growth in yield of major crops in the countries of the region has been much lower than the growth in value of agricultural outputs, it can be inferred that there is value addition happening through livestock farming and dairy farming, which are endemic to the region.

7.3 Findings and conclusions

Agricultural productivity has increased dramatically in South Asia over the past four to five decades, with Green revolution marking its beginning. The high population growth in the region had necessitated increased agricultural outputs, especially outputs of cereals, to meet the growing challenge of food shortage. The three important factors that had contributed to the growth are the increase in yield of major cereals, increase in cropping intensity resulting from the expansion in irrigation facilities, and to a lesser extent by arable land expansion. Among these three factors, arable land expansion had the smallest effect, due to the limited possibility of growth of cultivated land in the

[h] The total groundwater draft in South Asia can be anywhere between 300 BCM and 320 BCM, whereas the over-draft, which is mostly concentrated in western and north western India, could be in the range of 3000–4000 MCM, i.e., 1–1.2%.

region. Intensification of cropping had the second-largest effect, which again was possible with the creation of irrigation facilities, especially wells. Though the effect of yield improvement was the largest, it is quite likely that in many areas, the yield improvement witnessed is either because of the realization of genetic yield potential with the crop water requirement is fully met by the provision of irrigation and fertilizers or by virtue of more farmers adopting high yielding varieties with better inputs due to the availability of irrigation water.

While aggregate datasets suggest a 100% improvement in crop water productivity at the global level during the 40-year period (1961–2001) due to yield growth (source: http://www.fao.org/3/y4525e/y4525e06.htm), the effect of yield growth achieved for cereals in South Asia over the years on water productivity for those crops is not fully known. This is because there is insufficient information on the change in consumptive use of water (ET) for such crops occurring as a result of irrigation. However, with the increase in irrigation inputs, the yield and net income from crops had increased substantially, disproportionately higher than the extent to which crop consumptive use in the form of ET had increased, as evidence suggests. Studies in the Narmada river basin in India showed that the average water productivity for irrigated crops is far greater than the water productivity of rainfed crops (Kumar, 2010). This means, with the increasing volume of water from a basin being appropriated for irrigated crop production, the overall water productivity in agriculture in the basin improves considerably. However, this trend cannot continue once the basins are "closed" when every drop of water generated in the basin annually is used and reused, except in exceptionally wet years. The opportunity for further improvement in water productivity will have to come from reduction in nonbeneficial evaporation and nonrecoverable deep percolation.

The increasing adoption of microirrigation systems, particularly drip irrigation and mulching for row crops in India witnessed in the naturally water-scarce regions of India helps reduce nonbeneficial consumptive and nonconsumptive uses of water. Mulching combined with drip irrigation can convert the nonbeneficial consumptive use into beneficial transpiration, apart from reducing nonrecoverable deep percolation (nonbeneficial nonconsumptive use). With the area under drip irrigation alone exceeding 3.14 m. ha in the Indian states that are known for water scarcity (in 2016–2017), some notable improvement in agricultural water productivity must have been witnessed. Another notable change witnessed in the agricultural sector of South Asia is crop diversification. In the recent past, the region has also witnessed considerably high growth in agricultural output in value terms due to the shift to high-value products such as fruits, vegetables, and milk (Joshi et al., 2004).

References

Abdullaev, I., & Molden, D. (2004). Spatial and temporal variability of water productivity in the Syr Darya Basin, central Asia. *Water Resources Research, 40*(8).

Alexandratos, N., & Bruinsma, J. (2012). World agriculture towards 2030/2050: The 2012 revision. ESA Working Paper # 12–03, Agricultural Economics Development Division, Food and Agriculture Organization of the United Nations, June 2012.

Allen, R. G., Willardson, L. S., & Frederiksen, H. (1997). Water use definitions and their use for assessing the impacts of water conservation. In M. de Jager, L. P. Vermes, & R. Rageb (Eds.), *Sustainable irrigation in areas of water scarcity and drought* (pp. 72–82). England: Oxford.

Barker, R., & Molle, F. (2004). Evolution of irrigation in South and Southeast Asia. Comprehensive assessment of water management in agriculture (No. 5). Research report.

Deng, X. P., Shan, L., Zhang, H., & Turner, N. C. (2006). Improving agricultural water use efficiency in arid and semiarid areas of China. *Agricultural Water Management, 80*(1–3), 23–40.

Fischer, G., Hizsnyik, E., Prieler, S., & Wiberg, D. (2011). *Scarcity and abundance of land resources: Competing uses and the shrinking land resource base.*

Fischer, G., van Velthuizen, H., & Nachtergaele, F. (2009). *Global agro-ecological assessment-the 2009 revision.*

Food and Agricultural Organization of the United Nations. (2003). *Unlocking the value water in agriculture.* Rome: FAO.

Government of Pakistan. (2012). *Canal water pricing for irrigation in Pakistan: Assessment, issues and options.* Planning Commission, Government of Pakistan.

Jagadeesan, S., & Kumar, M. D. (2015). *The Sardar Sarovar Project: Assessing economic and social impacts.* New Delhi: SAGE Publications.

Joshi, P. K., Gulati, A., Birthal, P. S., & Tewari, L. (2004). Agriculture diversification in South Asia: Patterns, determinants and policy implications. *Economic and Political Weekly,* 2457–2467.

Kumar, M. D. (2010). *Managing water in river basins: Hydrology, economics, and institutions.* Oxford University Press.

Kumar, M. D., Patel, A., Ravindranath, R., & Singh, O. P. (2008). Chasing a mirage: Water harvesting and artificial recharge in naturally water-scarce regions. *Economic and Political Weekly,* 61–71.

Kumar, M. D., & Perry, C. J. (2019). What can explain groundwater rejuvenation in Gujarat in recent years? *International Journal of Water Resources Development, 35*(5), 891–906.

Kumar, M. D., Singh, O. P., & Sivamohan, M. V. K. (2010). Have diesel price hikes actually led to farmer distress in India? *Water International, 35*(3), 270–284.

Kumar, M. D., Sivamohan, M. V. K., & Narayanamoorthy, A. (2012). The food security challenge of the food-land-water nexus in India. *Food Security, 4*(4), 539–556.

Kumar, M. D., Turral, H., Sharma, B., Amarasinghe, U., & Singh, O. P. (2008). Water saving and yield enhancing micro-irrigation technologies in India when and where can they become best bet technologies? In *Proceedings of the 7th Annual Partners' Meet of IWMI-Tata water policy research program.* Hyderabad: International Water Management Institute, South Asia Regional Office, ICRISAT Campus.

Kumar, M. D., & van Dam, J. C. (2013). Drivers of change in agricultural water productivity and its improvement at basin scale in developing economies. *Water International, 38*(3), 312–325.

Kumar, M. D., & Vedantam, N. (2016). Groundwater use and decline in tank irrigation? Analysis from erstwhile Andhra Pradesh. In M. D. Kumar, Y. Kabir, & A. J. James (Eds.), *Rural water systems for multiple uses and livelihood security*. Elsevier.

Kumar, M. D., Viswanathan, P. K., & Bassi, N. (2014). Water scarcity and pollution in south and Southeast Asia: Problems and challenges (Chapter 12). In P. G. Harris, & G. Lang (Eds.), *Routledge handbook of environment and society in Asia* (pp. 197–215). Taylor & Francis Group, London: Routledge.

Ladha, J., Fischer, K., Hossain, M., Hobbs, P., & Hardy, B. (2000). *Improving the productivity and sustainability of rice-wheat systems of the Indo-Gangetic plains: A synthesis of NARS-IRRI partnership research*.

Mekonnen, M. M., & Hoekstra, A. (2012). A global assessment of the water footprint of farm animal products. *Ecosystems, 15*(3), 401–415.

Molden, D., Murray-Rust, H., Sakthivadivel, R., & Makin, I. (2003). A water-productivity framework for understanding and action. In J. W. Kijne, R. Barker, & D. J. Molden (Eds.), *Water productivity in agriculture: Limits and opportunities for improvement* (Vol. 1). CABI Publishing.

Pathak, H., Ladha, J. K., Aggarwal, P. K., Peng, S., Das, S., Singh, Y., … Aggarwal, H. P. (2003). Trends of climatic potential and on-farm yields of rice and wheat in the Indo-Gangetic Plains. *Field Crops Research, 80*(3), 223–234.

Pingali, P. L. (1999). *CIMMYT 1998–99 world wheat facts and trends. Global wheat research in a changing world: Challenges and achievements*. (No. 557-2016-38818).

Pingali, P. L. (2012). Green Revolution: Impacts, limits, and the path ahead. *PNAS, 109*(31), 12302–12308.

Qureshi, A. S., Ahmed, Z., & Krupnik, T. J. (2014). Groundwater management in Bangladesh: An analysis of problems and opportunities. In *Cereal Systems Initiative for South Asia Mechanization and Irrigation (CSISA-MI) project, research report no. 2., Dhaka*. Bangladesh: CIMMYT.

Qureshi, A. S., McCornick, P. G., Sarwar, A., & Sharma, B. R. (2010). Challenges and prospects of sustainable groundwater management in the Indus Basin, Pakistan. *Water Resources Management, 24*(8), 1551–1569.

Rada, N. E. (2013). Agricultural growth in India: Examining the post-green revolution transition. In *Selected paper prepared for presentation at the Agricultural & Applied Economics Association's 2013 AAEA & CAES Joint Annual Meeting. Washington D.C., August 4–6, 2013*.

Schmidhalter, U., & Oertli, J. J. (1991). Transpiration/biomass ratio for carrots as affected by salinity, nutrient supply and soil aeration. *Plant and Soil, 135*, 125–132.

Siddique, K. H. M., Loss, S. P., Herwig, S. P., & Wilson, J. M. (1996). Growth, yield and neurotoxin (ODAP) concentration of three Lathyrus species in Mediterranean-type environments of Western Australia. *Australian Journal of Experimental Agriculture, 36*(2), 209–218.

Siddique, K. H. M., Tennant, D., Perry, M. W., & Belford, R. K. (1990). Water use and water use efficiency of old and modern wheat cultivars in a Mediterranean-type environment. *Australian Journal of Agricultural Research, 41*(3), 431–447.

Siebert, S., Portmann, F. T., & Döll, P. (2010). Global patterns of cropland use intensity. *Remote Sensing, 2*(7), 1625–1643.

Talati, J., & Kumar, M. D. (2019). From watershed to sub-basins: Analyzing the scale effects of watershed development in Narmada river basin, Central India. In M. D. Kumar, V. R. Reddy, & A. J. James (Eds.), *From catchment management to managing river basins: Science, technology choices, institutions and policy*. Amsterdam, Netherlands: Elsevier Science.

Timsiha, J., & Connor, D. J. (2001). Productivity and management of rice-wheat cropping systems: Issues and challenges. *Field Crop Research, 69*, 93–132.

Tran, D. V. (1997). World rice production: Main issues and technical possibilities. In Chataigner J. (Ed.), *Activités de recherche sur le riz en climat méditerranéen* (pp. 57–69). Montpellier: CIHEAM (Cahiers Option s Méditerranéennes; n. 2 4 (2)).

Vos, R. (2018). *Agricultural and rural transformations in Asian development: Past trends and future challenges (No. 2018/87).* WIDER Working Paper.

Zhang, H. (2003). Improving water productivity through deficit irrigation: Examples from Syria, North China Plain and Oregon, USA. In J. W. Kijne, R. Barker, & D. J. Molden (Eds.), *Water productivity in agriculture: Limits and opportunities for improvement* (Vol. 1). CABI Publishing.

Zwart, S. J., & Bastiaanssen, W. G. (2004). Review of measured crop water productivity values for irrigated wheat, rice, cotton and maize. *Agricultural Water Management, 69*(2), 115–133.

CHAPTER 8

Measure for raising crop water productivity in South Asia and Sub-Saharan Africa

8.1 Introduction

South Asia is one of the fastest-growing regions in the world in terms of both population growth and economic development. The (compounded) annual population growth rate of the region is 1.19%, with significant variation across countries from a lowest of −0.07 in the Maldives to a highest of 2.34% in Afghanistan.[a] On the economic front, the region's GDP (at constant prices) has grown from 364.6 billion dollars in 1979 to 3.571 trillion dollars in 2019. In spite of the high population growth experienced in the past, the per capita GDP has grown from US$ 414.56 to US$ 1945.37 during this period. Both these drivers are expected to have a remarkable impact on the demand for agricultural commodities, though remarkable intraregional disparity exists. While the impact of the former on demand for agricultural commodities can be linear, the impact of the latter will be somewhat exponential, as the experience of developed countries and China and India suggest.

Many countries of South Asia (Pakistan and Afghanistan) and also large regions within some countries (like India and Sri Lanka) experience acute scarcity of water for growing crops. In countries like India, there is increasing pressure to reallocate water from agriculture to other sectors of the economy, as urbanization and industrialization happen rapidly. The pressure to find water for the environment is also high. The demand for raising water productivity in agriculture should be a priority in these countries, even if it

[a] The annual compounded growth rate of population for the rest of the countries is as follows. It is 1.19% for India; 1.45% for Pakistan; 1.25% for Nepal, 0.8% for Sri Lanka; 1.05% for Bangladesh; and 1.09% for Burma.

means reducing yield per unit area in certain cases where land is relatively abundant (Kumar & van Dam, 2013), while opportunities exist for increasing the effective utilization of water through the transfer of water from water-abundant regions to water-scarce regions. At the same time, in large parts of Bangladesh, one of the most populated countries of the subcontinent, water is abundant, but the pressure on land is huge due to very high population density. Raising crop yields should be the priority in that country even if it means low water productivity as land, not water, is the limiting factor for enhancing production. There are large hilly and mountainous regions in South Asia (northeastern parts of India, and large parts of Nepal and Bhutan) and cold mountain areas in India, Pakistan, and Afghanistan. These regions experience water shortage during a few months of the year and therefore harvesting water in the hills for lean season use is important as maintaining high efficiency of use of water.

The next section of this chapter discusses the heterogeneous water ecology of South Asia, in a way that provides a rapid assessment of the opportunities and challenges the regions offer for agricultural productivity growth. The subsequent section deals with the strategies for enhancing agricultural productivity in the region, covering the specific interventions for improving the utilization of water resources for crop production, technical efficiency improvements, other measures for improving water productivity in irrigated crops, and the institutional and policy measures. The fourth section deals with the interventions in rainfed agriculture in distinct water ecologies of South Asia.

8.2 The heterogeneous water ecology of South Asia

South Asia comprises eight countries viz. Afghanistan, Bangladesh, Bhutan, India, the Maldives, Nepal, Pakistan, and Sri Lanka. The region has a population of about 1.5 billion (22% of the world's population) and is the most populated region in the world. But it has only 4.8% of the world's total land area (Lal, 2005). Among different south Asian countries, India is the largest with about two-thirds of the geographical area and coastline, and nearly three-quarters of the population. Its topography includes a variety of mountains, plateaus, dry regions, intervening structural basins, and beaches. The elevation varies from the world's highest point, Mount Everest, to the world's lowest, the sea beach. It has about 10,000 km of coastline. The region has a largely tropical monsoon climate with two monsoon systems: the southwest monsoon (June to September) and the northeast monsoon (December to April). The region features large year-to-year variations in rainfall which frequently cause severe floods and droughts over large areas. South Asia has some of the world's largest river systems: the Indus river flows from China to Pakistan, the Ganga, and Brahmaputra flow for about 2525 and 2900 km, respectively, through Tibet, India, and Bangladesh (Sharda, 2011).

The region is characterized by diverse climates, and equally diverse soil and water resources. The region shows extraordinarily diverse landforms due to the diverse climatic regimes, latitudes, altitudes, and topography. Afghanistan and Bhutan are mostly rugged with mountains. Bangladesh is mainly flat alluvial plains. India has an upland plain (Deccan Plateau) in the south, a flat to rolling plain along the Ganges, deserts in the west, and the mountainous Himalayas in the north. The topography in the Maldives is flat with white sandy beaches. Nepal has the flat river plain of the Ganges in the south, a central hill region, and the rugged Himalayas in the north. Pakistan has the flat Indus plain in the east, mountains in the north and northwest, and the Baluchistan plateau in the west. The terrain of Sri Lanka is mostly low, flat to rolling plains with mountains in the south-central interior. Land degradation is one of the biggest problems in South Asia due to water erosion resulting from the steep topography coupled with high-intensity rainfall. Modern methods of agriculture further aggravate the situation, with practices such as overuse of fertilizers and pesticides, excessive irrigation of saline lands, and shifting cultivation (Rao et al., 2016).

South Asia is the most intensively cultivated region in the world (Siebert et al., 2010). Agricultural land occupies more than 50% of the land area in Afghanistan, Bangladesh, and India and less than 50% in the other countries. Most of the South Asian region is under rainfed agriculture. Afghanistan, Bhutan, and Sri Lanka are predominantly rainfed (≥80%), as are India and Nepal (60%–70%). Irrigated agriculture predominates in Pakistan (26%) and Bangladesh (45%). India, Bangladesh, Pakistan, Nepal, and Sri Lanka produce a wide range of agricultural and animal husbandry products (Rao et al., 2016).

The climate in Afghanistan is arid to semiarid. Mountains in Afghanistan cause many variations in climate. More than three-quarters of the annual precipitation (327 mm) is received as snow in the mountain ranges of central Afghanistan, and the stream flows in the river's peak from March to June. Bangladesh is located in the deltaic plains of river basins and the seashore. This country has a tropical climate—summer (March to June) is hot and humid while winter (October to March) is mild—the rainy season is (June to October) warm and humid. Annual rainfall is more than 2500 mm. Bhutan has a tropical climate in the southern plains, cool winters and hot summers in the central valleys, and severe winters and cool summers in the

Himalayas. Annual precipitation is 2200 mm. The Indian climate varies from tropical in the south to temperate in the north. The annual average precipitation in India is 1083 mm with about 85% of this rainfall received during the southwest monsoon. The Maldives has a tropical; hot and humid climate. Annual precipitation is about 2000 mm. The climate in Nepal varies from cool summers and severe winters in the north to subtropical summers and mild winters in the south. Annual precipitation is about 1500 mm. Pakistan's climate is mostly hot, desert with a temperate northwest and arctic north. Annual precipitation is about 500 mm (Rao et al., 2016).

Sri Lanka is an island nation with a tropical monsoon. The average annual rainfall is around 1700 mm. The northeast monsoon occurs between December and March and the southwest monsoon between June and October. Droughts affect the economies of South Asia in general, threatening food security in particular. Floods occur in India, Bangladesh, Pakistan, Nepal, and Afghanistan and can cause substantial damage to standing crops. Cyclones and severe thunderstorms cause irrevocable damage to the general life of the public as well as to agriculture and livestock. Bangladesh is extremely prone to floods and cyclones. The westernmost and easternmost parts of Bangladesh are drought-prone. About 33% of India receives less than 750-mm rainfall, and 68% of the sown area is subject to drought in varying degrees. Floods and cyclones are also frequent in India. The east coast of the country is hit by more cyclones. Sri Lanka has been experiencing drought since ancient times. An average of 11,000 ha of paddy land is destroyed every year due to the lack of water in sufficient quantities (Bhaskara Rao, 2011). Per capita, renewable water resources are highest in Bhutan due to its low population. Bangladesh and Nepal have reasonably good water resources on a per capita basis. Per capita, water resources in the remaining five countries range from 1000 to 2500 m^3 per year (Rao et al., 2016).

8.3 Water ecology of Sub-Saharan Africa[b]

8.3.1 Water resources of the region

In general, Sub-Saharan Africa has a low average runoff per unit area and high evaporation, and as a consequence of these factors, has less runoff discharging into the sea per unit area than most other regions of the world. The Nile discharges into the Mediterranean, ten into the Indian Ocean, three

---

[b] This section of the chapter (8.3) draws heavily from FAO (1986): Irrigation in Africa south of the Sahara, FAO Investment Centre Technical Paper 5, Rome, Italy.

into the South Atlantic, and four into the North Atlantic. Two major rivers, the Niger and the Nile passing through major inland swamps, where they lose a considerable amount of water through evaporation. Several other basins-most notably the Lake Chad system—are fully landlocked and eventually lose all their water by evaporation or infiltration.

With the exception of the Congo, all the major rivers of SSA show considerable seasonal variation inflow. This is most marked where there are no natural regulators such as lakes or swamps, or in rivers draining savannah or semiarid areas, with their possibilities of intense short-term rainfall. Natural sediment loads in African rivers are lower than in other regions.[c]

Surface water is very unevenly distributed. The Congo Basin, which occupies some 16% of the surface of Sub-Saharan Africa has 55% of the mean annual water discharge. A further seven rivers (the Niger, Ogooue, Zambezi, Nile, Sanga, Chari-Lagone, and Volta) contribute a further 25%. Only a few major rivers—most notably the Senegal, Niger, and Nile—flow through the substantial drought-prone areas mentioned earlier, where there are severe climatic restrictions on rainfed agriculture.

Detailed studies of groundwater potential in the region have covered only limited areas. These studies have suggested that: (i) the rocks of the basement complex which underlie over a half of SSA contain only small, discontinuous aquifers. Recharge rates are usually too slow and the water is often also too deep; (ii) groundwater yields in some sedimentary basins are generally intermediate, although in some cases (e.g., Somalia and Sudan) the water may be over 100-m deep and hence very costly to extract for irrigation; and (iii) shallow reserves of groundwater are most likely to be found in quantities along the alluvial beds of some of the major rivers—for instance in Northern Nigeria and coastal deltas and plains. In a few places—e.g., parts of Mali—artesian groundwater is known to exist.

More intensive recent works have identified zones in the Sahelian countries, comprising between 10% and 30% of their areas, which can yield shallow groundwater in volumes adequate for localized small-scale irrigation. Similar conditions appear to exist in southern Sudan and the drier parts of East Africa.

---

[c] However, despite this general picture they can be locally high in the rapidly increasing areas which have lost their vegetation cover due to human activity, either from overgrazing or deforestation. Mainly for this reason many of the smaller reservoirs constructed in the region were silting up faster than expected (FAO, 1986).

8.3.2 Agro climate of the region

Sub-Saharan Africa has a great diversity of climates and soils. The topography of much of SSA consists of undulating or gently sloping plains, at a variety of altitudes. Typically, these relatively flat areas are bordered by sharply rising or falling land representing the eroded edges of ancient geological features, volcanoes, or in the case of the rift valley system of East Africa, the results of continental drift. Most of the coastline is without indentations or a broad coastal plain; there are no major coastal deltas.

Most of Africa is geologically ancient and highly weathered and is often based on crystalline rocks. Soils tend to be leached and deficient in major nutrients, and also lack organic matter due to high temperatures. Often, they have only limited potential for arable agriculture. There are, however, patches of more fertile alluvium along with the present and past courses of major rivers, as well as inherently fertile vertisols deposited in old lake beds—the black cracking soils of the Sudan Gezira, are the most extensive example of the latter. Both riverine alluvium and vertisols have attracted irrigated agriculture. Irrigators have also been drawn to peaty soils formed under swampy conditions. In most cases, however, potentially irrigable soils are in scattered patches rather than large blocks, and often at some distance from a potential water source.

There are huge variations in climate. Mean annual rainfall ranges from a few millimeters in the central Sahara to several meters in parts of the humid tropical zone of West Africa. Potential evapotranspiration ranges from under 1500 mm per year in more humid areas from 2000 to 2500 mm per year in semiarid and arid zones. Cloud cover is relatively low in much of SSA for much of the year. Most of the region has temperatures that permit year-round crop growth but at higher altitudes and in the southernmost countries, there may be temperature limitations, and sometimes a frost risk. In semiarid and arid areas, high temperatures may be a limiting factor; but even in the Sahelian belt of West Africa, there are seasonally lower temperatures and cloud cover which are sufficient to limit yields of rice, the most common irrigated crop.

The climate of SSA can be very broadly divided into three major zones.

(i) The semiarid/Sahelian zone: The mean annual rainfall is below 800 mm and may be less than 100 mm, and is usually very erratically distributed within the rainy season. Up to 20% of the annual rainfall may occur within a single day, giving very heavy runoff and a high potential for erosion. The growing season for annual crops is below 100 days in much of the zone. Irrigation is essential for crop production. Low-intensity grazing is the traditionally preferred land use. Many countries span two or all three of these climatic zones.

(ii) The savannah zone: The mean annual rainfall is between about 800 and 1200 mm. Within a season, rainfall patterns, as well as seasonal totals, may be erratic, with as much as a tenth of the seasonal total falling in 1 day; the growing period for annual crops ranges from 120 to 240 days. In areas with increased risks of crop failure due to drought, the length of the growing period is below 200 days. Supplementary irrigation in this region can give worthwhile benefits even in average rainy seasons, by compensating for within-season dry spells. Irrigation is essential for dry season annual cropping and many perennial crops.

(iii) The humid tropical zone: Mean annual rainfall exceeds 1200 mm and is usually over 1500 mm. Rain is fairly well distributed, seldom with more than four dry months. The growing period for annual crops usually exceeds 280 days per year. Irrigation is seldom economically justified except for dry-season supplementation of some perennial crops.

Many countries of SS Africa have two or all three of these climatic zones.

Long crop-growing periods under rainfed conditions indicate adequate soil moisture due to sufficient precipitation and low evaporation. That said, there are eight countries of SS Africa that have little or no land with a rainfed growing period above 200 days. They are Burkina Faso, Mali, Somalia, Niger, Senegal, Mauritania, Botswana, and Kenya. For these countries, irrigation is likely to be an essential part of any overall national strategy for increased agricultural production. Further, there are 12 countries that have a rainfed growing period of fewer than 120 days on over a quarter of their territory. They are Mauritania, Niger, Burkina Faso, Mali, Chad, Sudan, Ethiopia, Somalia, Kenya, Tanzania, Zimbabwe, and Botswana. Another 10 countries have up to a quarter of their area in the same semiarid zone.

8.4 How to enhance agricultural water productivity in South Asia

The demand for agricultural commodities in South Asia is expected to increase at an average rate of 1.60% per annum in the coming decades. From the earlier section, it is evident that most of this increased demand will have to be met through yield growth and expansion in the irrigated areas since not many additional land resources are available in the region for expansion. Yield growth can come from the introduction of fast-growing, high-yielding varieties, and also from improved crop management, including supplementary irrigation to the crops that are rain-fed. While the former improves water productivity (both at field level and regional level), with no impact on water demand, the latter case will increase water demand in agriculture. But, the last option, i.e., of increasing irrigation intensity (the ratio of total irrigated area and the cultivable land) will be possible only through the exploitation of additional water resources or through improvements in water productivity of consumed water. If additional water resources are not available for exploitation, CWP will have to be improved at the rate at which demand for agricultural commodities is likely to increase in the future, either through yield improvement or through improvement in technical efficiency of water use (reducing soil evaporation and other nonbeneficial consumptive uses) or both.

Despite the increases in land under cultivation in the land-abundant countries, much of agricultural productivity growth has been based on the growth of yields, and will increasingly need to be so. What is the potential for a continuation of yield growth? In countries and localities where the potential of existing technology is being exploited fully, subject to the agroecological constraints specific to each locality, further growth, or even maintenance, of current yield levels will depend crucially on further progress in agricultural research. In places where yields are nearing ceilings obtained on research stations, the scope for raising yields is much more limited than in the past (Sinclair, 1998). Despite this, average yields have continued to increase, albeit at a decelerating rate. But the divergence between economically efficient and agroecologically attainable yields can be very wide (Bruinsma, 2009).[d]

We will explore the potential for yield improvements, and intensifying irrigation and water productivity improvements in the context of South Asian countries, and the role of technologies, policies, and institutions in achieving these in the subsequent sections.

---

[d] For example, the United Kingdom and the United States of America have nearly equal attainable yields (6.0–6.3 ton/ha) but actual yields are 7.8 ton/ha in the United Kingdom, exceeding the attainable yields suggested by GAEZ evaluation) and 2.8 ton/ha in the US. In spite of US yields being a small percentage of those that are agro-ecologically attainable and of those prevailing in the UK, it is not necessarily a less efficient wheat producer than the UK in terms of production costs (Sinclair, 1998).

8.4.1 Increasing the utilization potential of the available water resources

As discussed in the previous section, the scope for utilization of the renewable water resources in the water surplus basins of South Asia within the basin for enhancing beneficial evapotranspiration is extremely low. The constraint is induced by a lack of a sufficient amount of arable land for cultivation, agroecological constraints such as floods, waterlogging and salinity, and lack of sunshine. In certain cases, irrigation would become too expensive due to the amount of energy required for lifting the water from the points of storage to the places where the actual demand for water exists. In the water-scarce basins, the surface water resources are already over-appropriated and groundwater resources have depleted, with mining in some localities. Greater utilization of the surplus water from these basins (which now drains into natural sinks) would be possible only if the excess flow is transferred to basins, which are water-scarce.

While building regional cooperation for such transfers between a water-rich country and a water-scarce region will be a distant possibility, within countries such possibilities exist. The operationalizing of such concepts is ideal for countries that are characterized by interregional variations in demand–supply balance for water resources. India is one of them, wherein the regions which are rich in water resources have very poor demand for water, and regions that have limited water resources have high water demand (Kumar, Vedantam, Puri, & Bassi, 2012). India has already conceptualized a national river linking plan which involves several links to transfer water among basins within the Himalayan river systems and among basins within the peninsular river systems.

This is one large-scale intervention that will have major impacts on agricultural production, productivity and can lead to enhanced regional water productivity in agriculture, as a portion of the large volume of water that now goes uncaptured into the natural sinks (oceans), from the Himalayan rivers (the Ganges, Brahmaputra, and Meghna) and a few peninsular Indian rivers (Godavari) would get utilized for expanding irrigation in the water-scarce regions of western and southern India.

Following are the changes that would drive water productivity improvements. At the field/plot level, increased availability of a dependable source of water would encourage replacement of rain-fed, short-duration varieties of crops with low yields by their long-duration, high-yielding irrigated counterparts, with a resultant increase in ET, yield, and water productivity. At the regional level, increased irrigation water availability would also lead to an increase in irrigation intensity, with the greater amount of blue water being utilized for beneficial ET. But there will be no expansion of arable land in India. Hence, yield enhancement and cropping intensity would drive crop production growth, with a major improvement in water productivity at the regional level.

In Afghanistan, the existing (surface) irrigation systems are not properly utilized as they lie in dilapidated conditions after several years of war and conflicts (Qureshi, 2002). The rehabilitation of these systems, if supported by improved security in the region, can lead to better utilization of the irrigation infrastructure for bringing in the available arable land under cultivation. This can lead to an improvement in system-level and regional-level water productivity.

8.4.2 Technology changes for technical efficiency improvements

Technology changes for enhancing water productivity would mainly include: (i) crop technologies with improved genetic yield potential and increased resistance to stresses, particularly droughts, waterlogging and diseases; (ii) reducing the yield gap, through optimum use of a wide range of inputs viz., irrigation, fertilizer, labor, greater adoption of improved varieties, improved farm management practices including weed control, pest control and water control; (iii) improved irrigation technologies which improve water use efficiency through raising yields and reducing consumptive use of water[e]; and (iv) improved fertigation techniques, which improve the nutrient use efficiency.

As regards the genetic yield potential, the focus should be on crops viz., rice and maize, and to an extent, wheat. At the next level, the focus should be on reducing the yield gap—the gap between the average yield realized by the farmers and the best yield obtained by farmers, or the yield which is economically viable in a particular region. Since one of the primary reasons for the yield gap is yield loss due to stresses caused by disease, droughts, waterlogging, and salinity, this to an extent can be addressed through the introduction of stress-resistant varieties. In the context of South Asia, this aspect is particularly important as vast regions in the subcontinent are vulnerable to droughts (paddy-wheat growing regions of Punjab and Sindh in Pakistan,

---

[e] Consumptive use here does not refer to ET, but ET plus nonbeneficial evaporation from barren soil and the nonrecoverable deep percolation (see Allen, Willardson, & Frederiksen, 1997 for details of the terminology).

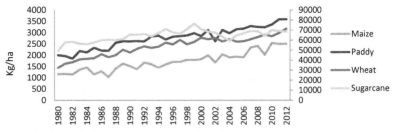

Fig. 8.1 Yield trends for major crops in India; 1980–2012.

paddy-wheat growing areas of northwestern India, wheat-growing areas of western India, paddy growing areas of peninsular India), floods (paddy growing areas of eastern India, Bangladesh).

8.4.2.1 Reducing yield gaps

As regards the yield gap, though the yields for the four major crops have been rising across South Asia (Fig. 8.1 for India (based on FAOSTAT for the period from 1980 to 2012), and Table 7.2 of Chapter 7 for South Asian countries), the rate of growth has been different in different regions. The yield gap is very high for major cereals viz., wheat, and paddy, in the Eastern Gangetic plains in India (Eastern UP, Bihar) and Bangladesh. There are many factors contributing to this. First is agroecological—floods and water-logging. The second is poor crop management. The third is related to the adoption of varieties having high genetic potential. But all these issues are interrelated. As noted by Pingali and Helsey (1999) and Byerlee and Traxler (1996), in environments favorable to wheat production, the gap between economically exploitable and the yields achieved on farmers' fields has been reduced considerably in the past. This means, in these areas the cost of marginal increments in yield, could exceed the incremental gain. The cost is high not only in terms of increased use of inputs such as fertilizer, fuel, and water but also in terms of increased management and supervision time for achieving more efficient input use.

The yield gap is very high in rice. Rice in India is grown almost throughout the country except in the arid parts of Rajasthan. It is grown in extremely diverse hydrological environments such as irrigated, rain-fed uplands, lowlands, as well as under deep-water conditions. Of the 44 m per ha of harvested area, almost 54% is irrigated. Most of the rice-producing areas of Punjab, Haryana, Andhra Pradesh, and Tamil Nadu are irrigated. Rain-fed rice is grown in states such as West Bengal, Uttar Pradesh, Orissa,

Bihar, Assam, Karnataka, Maharashtra, Madhya Pradesh, and Jharkhand, which receive high rainfall (Aggarwal et al., 2008).

The rice yields in India increased from around 2000 kg per ha in 1980 to 3591 kg per ha in 2012 mainly as a result of irrigation and the introduction of high-yielding varieties. The average yield of rice is more in several districts of Punjab, Haryana, Andhra Pradesh, and Tamil Nadu. The yields are generally less in central Indian states such as Madhya Pradesh and eastern Indian states such as Orissa, Bihar, and Jharkhand. Since rice is mostly rain-fed in these states, the production is strongly dependent on the distribution of rainfall (Aggarwal et al., 2008). Though the yield gaps for rain-fed paddy in relation to on-farm yield potential were found to be very high for states, viz., Madhya Pradesh, Uttar Pradesh, West Bengal, Orissa, and Jharkhand, most of it is caused by delayed transplanting, erratic monsoon rains, and poor nutrient management. They all point to the criticality of irrigation in raising farm yields (Aggarwal et al., 2008). But, barring UP and WB, most of these regions experience a shortage of water during summer months, when rains fail or become erratic, to start transplanting.

Whereas in the high rainfall regions such as UP, coastal Orissa, north Bihar, Assam, and West Bengal in India and flood-prone areas of Bangladesh, excessive rainfall and floods make nutrient management difficult. Hence overall there is too little one can achieve in terms of reducing the yield gap in paddy, unless genetic traits such as "drought resistance," "salt tolerance," and "flood resistance" are infused in the rice gene.

8.4.2.2 Efficient irrigation technologies

Improved irrigation technologies for South Asia would mainly include drip irrigation systems and plastic mulching. They can raise crop yields if irrigation scheduling is done scientifically while saving water, thereby raising water productivity substantially (Singh, Singh, & BHATT, 2014; Xie, Wang, & Li, 2005; Zhang, Wei, Han, Ren, & Jia, 2020; Zhang, Zhang, Qi, & Li, 2018). As we have already discussed, nearly 30% of the cropped area suitable for drip irrigation in India (5.8 m per ha), where system adoption would lead to real water saving, is already under the technology (Kumar, Turral, Sharma, Amarasinghe, & Singh, 2008). The additional area that can be covered under the technology is around 3.90 m per ha. These estimates, however, do not include areas under crops receiving canal water, that otherwise are amenable to drip technology, but cannot be put under the same due to lack of pressure for running drips. But such areas also can be covered under pressurized irrigation systems like the drips and sprinklers, with intermediate storage systems and pressurizing pumps, provided farmers have sufficient land to spare for constructing the storage tanks. In a few regions such as Rajasthan, the adoption of pressurized sprinklers has picked up on a large scale in the recent past, with the government subsidizing intermediate storage systems for farmers to store water from canals (Amarasinghe, Bhaduri, Singh, Ojha, & Anand, 2008).

Different types of microirrigation systems that are amenable to different crops and cropping systems are available. While some are only technically feasible and economically viable for row crops and orchards, some are feasible and economically viable for field crops also. They can improve crop water productivity by reducing the nonbeneficial evaporation or nonrecoverable deep percolation in the field, resulting in total depletion or consumed fraction (CF).[f] Or they can increase the proportion of the beneficial fraction of the applied water that leads to improved crop yield. Nevertheless, in both cases, water productivity improvements come without a reduction in yields (for details, see Kumar, Singh, Samad, Purohit, & Didyala, 2008; Palanisami, Gemma, & Ranganathan, 2008). There are other conservation technologies such as zero-tillage and bed planting which can improve water productivity in wheat (Sikka, 2009).

Mulching can save more water than drips, by affecting a reduction in soil evaporation. It can also increase crop yields, as greater utilization of water in the soil profile for transpiration (Jin, Zhang, Sun, & Gao, 1999; Singh et al., 2014; Xie et al., 2005; Zhang et al., 2018; Zhang et al., 2020). In India, by the most conservative estimates, the potential area under the plastic mulch will be as high as that of water-saving microirrigation technologies, and its use would further improve the water use efficiency.[g] However, there are many field crops, for which mulching is being experimented with worldwide to improve water use efficiency.

Pakistan is a water-scarce country and water use efficiency in irrigation is extremely crucial to save the scarce water resources in the country. The country has vast areas under cotton and sugarcane in the Indus Basin Irrigation System (IBIS) command. As per the 1995/1996 data, a total of 3.2 m per ha of land is under crops that are amenable to drip irrigation, such as cotton, sugarcane, and fruits. The water demand (PET) for cotton and sugarcane in the IBIS were estimated to be in the range of 627 to 1161 mm for

---

[f] See Allen et al. (1997) for various definitions.
[g] Here again, we only consider the costs where a significant proportion of the land is exposed to direct solar radiation, for considerable time periods.

cotton, 11,278 to 1887 mm for sugarcane for the traditional method of irrigation. The total water demand for these two crops is substantial (Ullah, Habib, & Muhammad, 2001). The use of plastic mulching and drip irrigation could save a substantial amount of water in cotton, which is depleted under the traditional method of irrigation, in nonbeneficial evaporation from the soil exposed to solar radiation, and an extent, the water which goes into deep percolation. But drip irrigation will not be a good idea in areas where the return flows from gravity irrigation are significant enough to augment groundwater recharge and improve groundwater quality, and would be most desirable in areas where the return flows go into saline groundwater. However, its use for applying water to the plants in drops is necessary to make plastic mulch more effective. The policy measures like the one tried successfully in the IGNP command area in Rajasthan can work in Pakistan as well for the uptake of the technology.

8.4.3 Other measures for improving water productivity in irrigated crops

A detailed study of three river basins in India, viz., Narmada, Sabarmati, parts of Indus, and Ganges demonstrated that there are five major avenues for improving water productivity in irrigation crops, with other scholars sharing similar views (Kumar, Singh, et al., 2008). They are (1) water delivery control; (2) improving quality and reliability of irrigation water supplies (Kumar, Singh, et al., 2008; Palanisami et al., 2008); (3) optimizing the use of fertilizers; (4) use of microirrigation systems (Palanisami et al., 2008; Sikka, 2009); and (5) growing certain crops in regions where they secure high water productivity due to climatic advantage.

8.4.3.1 Water delivery control

Studies in some basins of South Asia show that great opportunities for improving water productivity (in physical terms (part of Godavari basin in Maharashtra, India) and economic terms (Narmada basin, India) exist through control over water delivery. Fig. 9 shows the relationship between irrigation dosage and water productivity for sugarcane, as per the analysis carried out in around 56 farms in Maharashtra. As Fig. 8.2 suggests, the marginal WP in sugarcane with an extra dosage of irrigation is negative. In that case, limiting irrigation dosage might give a higher net return per m^3 of water as well as yield per m^3 of water. But only those farmers whose irrigation corresponds to the descending part of the "yield response curve" would be interested in this strategy which increases crop water productivity.

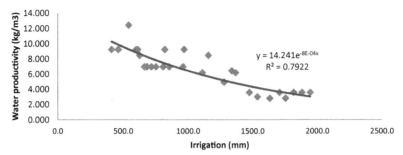

Fig. 8.2 Water productivity vs irrigation dosage: serpentine furrow. *Source: Kumar, M. D., Vedantam, N., Puri, S. & Bassi, N. (2012). Irrigation efficiencies and water productivity in sugarcane in Godavari River basin, Maharashtra, final report submitted to the world wild Fund for Nature, Institute for Resource Analysis and Policy, Hyderabad.*

For others, raising crop WP through this strategy means that they compromise on the yield and net return.

Even when the return from the land does not improve, the strategy of limiting irrigation can work through the following conditions: (a) the amount of water, farmers can access is limited by the natural environment; (b) farmers are confronted with high marginal costs of using water due to high prices for water or electricity used for groundwater pumping; (c) water supply is rationed; and (d) there is great social pressure from downstream farmers to improve efficiency, especially during droughts (Lankford, 2012). But, in all these situations, the farmers should have extra land for using the water saved (Kumar & van Dam, 2013). Within South Asia, the regions which have extra arable land for expanding irrigation are concentrated (thereby increasing cropping intensity) in peninsular India, western India, parts of central India, parts of the Indus basin area in Pakistan and Afghanistan.[h] These areas are also likely to experience the situation described above where irrigation corresponds to both ascending and descending parts of the yield response curve of irrigation.

However, the presence of extra land, the same factor, which creates incentives for raising WP, would force the smallholders of the South Asia region to redistribute the saved water for expanding irrigated area so as to sustain or enhance their farm income. The reason is that the amount of water being handled is so small that they would tend to use the same amount of water as done previously since the WP differences would be just marginal

[h] The per capita arable land is 0.25 ha in Afghanistan, 0.13 ha in India, 0.12 ha in Pakistan, 0.08 in Nepal, 0.06 in Sri Lanka, and 0.05 in Bangladesh (Kumar, Viswanathan, & Bassi, 2014).

(Kumar & van Dam, 2009). A significant increase in "basin water use efficiency" and economic output from farming at the basin level can be achieved through this, though there will no real water saving. Hence, the impact would be greater economic outputs for the same quantum of water (Kumar & van Dam, 2013).

Thus, it may appear that to affect demand reductions, it is important to ration water distribution in canals along with better education of the farmers about crop management. Proper regional and sectoral water allocation can become an incentive for improving WP. Experiences from the Murray–Darling basin (Haisman, 2005) and Chile (Thobanl, 1997) show significant improvements in WUE and value of water realized, respectively, in irrigated production after the introduction of volumetric rationing enforced through properly instituted water rights. Nevertheless, marginal water productivity analysis of the kind presented above can help decide on allocation and delivery strategies for canal water. Therefore, semiarid regions, which have a smaller percentage of their arable land under irrigated production, should resort to institutional measures for achieving the twin objective of water use efficiency improvement and water saving.

This means, allocating less water in many instances with a resultant reduction in yield but an increase in WP and allocating more water in certain instances with a resultant increase in both yield and WP (Kumar & van Dam, 2009) would help reduce water use if we decide not to increase crop outputs. Amarasinghe and Sharma (2009) showed conforming results. The study showed that water productivity in irrigated crops could be enhanced significantly through deficit irrigation, a key strategy in water delivery control in 251 districts. These are districts that already show a very high yield per unit of land, and receive intensive irrigation.

8.4.3.2 Taking advantage of climate to improve water productivity

Physical productivity of water is influenced by climatic factors—with solar radiation and temperature affecting yield, solar radiation, temperature, wind speed, and humidity affecting ET (Aggarwal, Kalra, Bandyopadhyay, & Selvarajan, 1995; Loomis & Connor, 1996; Ruthenberg, 1980) and agronomic factors—with crop variety affecting the potential yield and ET requirement (Hussain, Sakthivadivel, Amarasinghe, Mudasser, & Molden, 2003). Since yield affects the gross returns, the climate would have implications for water productivity in economic terms as well. Hence, certain crops give higher water productivity in both physical and economic terms by virtue of the climate under which they are grown without any additional inputs of nutrients and improved crop technology (Loomis & Connor, 1996: pp. 398). Studies in the Narmada basin show major differences in water productivity of wheat and irrigated paddy across nine agroclimatic subregions (Kumar, 2010).

In the case of wheat, the physical productivity of applied water for grain production during the normal year was estimated to be highest for the northern region of Chhattisgarh in Mandla district (1.80 kg per m^3) though Raisen falls in the traditional wheat-growing belt; it was lowest for Jabalpur in Central Narmada Valley (0.47 kg per m^3). This is mainly due to the major difference in irrigation water applied, which is 127 mm against 640 mm for Jabalpur. This is a significant difference, with the highest being 250% more than the lowest. The difference in irrigation can be attributed to the difference in climate between Jabalpur (dry semihumid) and Mandla (moist subhumid), which changes the crop water requirement. Higher biomass output per unit volume of water (physical productivity) should also result in higher economic output especially when the difference is mainly due to the climatic factors, which change the ET requirements unless the factors which determine the cost of inputs significantly differ. In this case here, it was found that the net economic return per cubic meter of water was highest for the same region for which physical productivity was higher (Rs. 4.09 m^3). But the same was lowest for Narsinghpur (Rs. 0.86 m^3), which had the second-lowest physical productivity.

As regards paddy, there are only two regions that irrigate paddy. The physical productivity for grain during the normal year was estimated to be higher for the Northern region of Chhattisgarh in the Mandla district (2.13 kg per m^3) whereas it was only 1.62 kg per m^3 in Jabalpur district of Central Narmada Valley. Likewise, the combined physical and economic efficiency of water use was found to be higher for Chhattisgarh (Rs. 3.59 per m^3) against Rs. 1.43 per m^3 for Jabalpur in Central Narmada Valley. Such climatic advantage exists in many major basins such as Indus, Ganges, Cauvery, Sabarmati, and Narmada with lower aridity, higher rainfall, and higher humidity experienced in the upper catchments (based on Kumar, Ghosh, Patel, Singh, & Ravindranath, 2006; Kumar, Patel, Ravindranath, & Singh, 2008). For instance, within the paddy-wheat growing area of Punjab, the climate varies from hot semiarid to hot and subhumid. This advantage can be tapped to allocate more land for water-intensive crops in localities where ET requirement is less and sunshine is more.

8.4.3.3 Crop shifts

Another major opportunity for water productivity improvement comes from crop shifts. In every region, the agro climate permits the growing of several different crops in the same season, and our analysis shows that there are major variations in water productivity in economic terms across crops (Kumar, 2010; Kumar, Singh, et al., 2008). Several of the cash crops such as castor, cotton, fennel, cumin, and ground nut and vegetable such as potato are found to have higher water productivity than the cereals grown in the same region (Kumar, 2010; Kumar, Singh, et al., 2008; Kumar & van Dam, 2009). But, if we consider the food security benefits of growing cereals, the opportunity available for WP improvement through crop shift may not be significant in major food-producing areas. In such areas, the opportunities for shifting from less water-efficient nonfood crops to water-efficient cash crops and fruits should be explored. Semiarid pockets such as north Gujarat, Saurashtra, central Madhya Pradesh, western Rajasthan, northern Karnataka, parts of Tamil Nadu, and western parts of AP are ideal for such crop shifts to improve crop water productivity and reduce the stress on groundwater. These are not major food-producing areas.

8.4.3.4 Crop management

There are many irrigated districts in eastern India which are dominated by food crops. The yield of food crops such as wheat and paddy is very low in these districts, yield gaps high (Pathak et al., 2003), and very low total factor productivity growth (Evenson, Pray, & Rosegrant, 1999). Amarasinghe and Sharma (2009) show that there are 202 districts in the country which fall under the category of medium consumptive use of water for irrigated crops (300–425 mm), but with high yield gaps. Improved agronomic inputs (high-yielding varieties and better use of fertilizers and pesticides) can significantly raise the yields. For instance, a meta-analysis involving regression of data from 54 individual cases of water use, yield, and water use efficiency variations in wheat and cotton showed a significant effect of crop management practices on WUE for both the crops, with effects of crop and soil management practices (viz., crop rotation, crop variety selection) and climate more significant in the case of cotton, and effect of microirrigation more significant in the case of wheat (Fan, Wang, & Nan, 2018). Such practices will have a positive impact on water productivity, though water productivity is not a concern for farmers in this water-abundant region of India. While there are districts in central India, where better use of fertilizers would help enhance crop yields, these areas also require an optimum dosage of irrigation to achieve this (Kumar, Singh, et al., 2008).

8.4.3.5 Improvement in quality and reliability of irrigation

As regards improvements in quality and reliability of irrigation, it is more relevant for canal irrigated areas, and areas receiving tank irrigation (Palanisami et al., 2008). The area irrigated by canals is high in Punjab, Haryana, Uttar Pradesh, Bihar, Maharashtra, Tamil Nadu, and Andhra Pradesh. Some of these areas have good native groundwater and farmers could supplement canal water with well water. Such areas include central and northeastern Punjab, Uttar Pradesh, and Bihar. Whereas in parts of Maharashtra, Tamil Nadu, and Andhra Pradesh, the hard rock aquifers get replenished due to return flows from canal irrigated fields and seepage from canals. This is already extensively practiced in Punjab, Maharashtra, and South Gujarat.

Sikka (2009) shows that introducing horticulture, fish and prawn farming, and duckery through secondary reservoirs and raised bed-cum-trench could enhance water productivity in economic terms in seasonally waterlogged areas of Bihar under the rice-wheat system in the orders of magnitude. Also, in seasonally water-logged paddy fields, the introduction of horticulture-fish production in raised bed cum trench was found to be economically viable. Many rice farms in eastern India are multiple-use systems with significant values being added by trees, fisheries, and dairying, and assessing their water productivity in relation to the returns from paddy production against the total water delivered would lead to significant underestimation of water productivity of such agricultural systems (Sikka, 2009). Nevertheless, the water accounting procedure adopted in the study did not take into account the increased water demand induced by trees or the actual amount of water directly used by trees from the subsurface strata. Hence, it is quite likely to have resulted in the over-estimation of incremental water productivity obtained under the farming system.

In some other areas, where groundwater is scarce or is of poor quality, the quality and reliability of irrigation water supplies could be improved through the creation of intermediate storage systems like the one found in the Bikaner district of Rajasthan. But, one prerequisite for this is the availability of land area for cultivation and farmers' ability to spare land for the construction of such storage systems. Area irrigated by tanks is high in the South Indian states of Andhra Pradesh and Tamil Nadu.

Kumar, Trivedi, and Singh (2009) developed a composite index for estimating irrigation quality and applied the same for three different sources of irrigation, viz., wells, canals, and well + canal in the Bist Doab area of Punjab in India to examine its impact on crop water productivity. The study found the irrigation water quality index to be higher for well-irrigated fields as compared with canal irrigated fields and fields irrigated by both wells and canals in one village. The same trend was found in another village (Changarwan), except for paddy, which was largely in conformation with the perception of farmers in vis-à-vis the quality of canal water received for paddy (See Table 8.1).

Further, differential reliability has an impact on the economic productivity of water (Rs. per m^3). The fields which received irrigation water of higher quality and reliability got higher water productivity in rupee terms. But the impact of differential quality and reliability was not visible on the physical productivity of water for fodder crops (Table 8.1). The findings of the research, however, contradicted the conventional wisdom that higher quality and reliability of irrigation would result in better yields. A deviation was found in one village, which was due to the differential chemical quality of water, which the index could not capture. This means that improved quality and reliability of irrigation would help enhance water productivity in crop production. The research also showed that the quality and reliability of irrigation water also had a significant impact on cropping patterns.

8.4.3.6 Improved crop management

Surprisingly, even in the most productive regions of South Asia, such as North-Western India, not all farmers achieve high levels of productivity in terms of the use of land and water. There is a wide variation in productivity between farmers, due to differences in transplanting and farm management. Singh, van Dam, and Feddes (2006) in their study in the Sirsa district of Haryana found considerable spatial variation in WP values for different crops but also the same crop. For instance, the WP_{ET}, expressed in terms of crop grain (or seed) yield per unit amount of evapotranspiration, varied from 1.22 to 1.56 kg m^{-3} for wheat among different farmer fields. The corresponding value for cotton varied from 0.09 to 0.31 kg m^{-3}. This indicates enormous scope exists for both improving yield and reducing the consumptive use of water for crop production through timely transplanting, improved nutrient management, weed and pest control, and controlled water delivery. The average WP_{ET} (kg m^{-3}) was 1.39 for wheat, 0.94 for rice, and 0.23 for cotton, and corresponds to average values for the climatic and growing conditions in Northwest India. Including percolation in the analysis, i.e., crop grain (or seed) yield per unit amount of evapotranspiration

Table 8.1 Irrigation water quality index and productivity of water for crops in Changarwan village.

Name of crop	Source of irrigation	Irrigation quality index	Irrigation water applied (m³/acre)	Crop yield (kg/acre)	Total input cost (Rs/acre)	Water productivity (kg/m³)	Water productivity (Rs/m³)
Maize	Well	10.28	598.7	941.7	1035.1	1.53	6.44
	Canal	0.65	2600.0	880.0	1135.1	0.53	2.00
Wheat	Well	2.26	915.4	1003.6	1723.7	1.97	4.45
	Canal	0.5	1109.0	1110.6	1889.9	1.57	3.46
Barseem	Well	0.44	1184.5	4864.6	1118.5	6.53	12.99
	Canal	0.17	2488.5	7216.7	1076.5	10.23	24.01
Paddy	Well	2.66	3518.5	1169.5	2180.6	0.57	0.32
	Canal	3.33	5849.8	1661.2	2065.4	0.41	1.50

Source: Based on Kumar, M.D., Trivedi, K., & Singh, O.P. (2009). Analyzing the impact of quality and reliability of irrigation water on crop water productivity using an irrigation quality index. In Kumar, M.D., & Amarasinghe, U. (Eds.). Water productivity improvements in Indian agriculture: Potentials, constraints and prospects. Strategic analyses of the National River Linking Project (NRLP) of India series 5, 163.

plus percolation, resulted in average WP_{ETQ} (kg m^{-3}) values of 1.04 for wheat, 0.84 for rice, and 0.21 for cotton.

Factors responsible for low WP include the relatively high evaporation ratio (E_s/ET) especially for rice, and percolation from field irrigations. Improving agronomic practices such as aerobic rice cultivation and soil mulching (for wheat and cotton) will reduce this nonbeneficial loss of water through evaporation, and subsequently improve the WP_{ET} at the field scale. For wheat, the simulated water and salt limited yields were 20%–60% higher than measured yields, which suggested substantial nutrition, pest, disease, and/or weed stresses. Singh et al. (2006) found that improved crop management in terms of timely sowing, optimum nutrient supply, and better pest, disease, and weed control for wheat will multiply its WP_{ET} by a factor of 1.5. Moreover, severe water stress was observed on cotton (relative transpiration <0.65) during the *Kharif* (summer) season, which resulted in 1.4–3.3 times lower water and salt limited yields compared with simulated potential yields.

8.4.4 Policy changes for both technical efficiency and allocative efficiency improvements

8.4.4.1 Irrigation water pricing

Both irrigation water and electricity are critical inputs for agriculture in South Asia. Largely, the existing mode of pricing of water and electricity for agriculture in South Asian countries is not efficient, as it does not induce the opportunity cost of using scarce resources such as water and electricity through price signals. There is sufficient empirical evidence to the effect that changes in the existing mode of pricing of water and electricity in the irrigation sector to "pro-rata" or consumption-based pricing can drive water productivity improvements in future in the region (Kumar, 2005; Kumar, 2010; Kumar, Scott, & Singh, 2011; Malik, 2008). But we need to examine where such policy instruments work. Currently, irrigation water is heavily subsidized in all South Asian countries, where public irrigation systems contribute significantly to national irrigation, be it India, Pakistan, Bangladesh, Sri Lanka, or Nepal. In Pakistan, the average price for canal water levied on a crop-area basis, is only 1/15th of that of the cost tube well owners incur for pumping groundwater in rupees per cubic meter of water. Further, the cess per ha of irrigation for a crop does not reflect the water requirement for the individual crop concerned (GOP, 2012). In India also, canal water pricing is on the basis of crop-area in all the states, though in a few canal commands, farmers have introduced volumetric water pricing under the participatory irrigation management program. Under such a pricing structure, farmers do not have the incentive to either use water efficiently in physical terms for individual crops or allocate the available water to crops that are efficient in economic terms.

But volumetric pricing of canal water is not an easy task. There are theoretical issues associated with fixing prices that would be socioeconomically viable while making it sufficiently high to induce scarcity value of the resource. It would also require infrastructural improvements in the water distribution and delivery system for measuring volumetric water delivery at various levels in the hydraulic system hierarchy for water control and measurements, and strong administrative mechanisms to detect illegal water tapping and penalize the payment defaulters (Perry, 2001). If introduced, volumetric pricing can impact water productivity in two ways. First, it would encourage farmers to apply water more efficiently to crops, reducing field "losses" such as runoff and deep percolation to reduce the cost of irrigation water (Kumar and Singh, 2001; Kumar, 2010), resulting in higher physical productivity of water for the existing crops.

Second, it would also encourage them to allocate more water to crops that yield very high returns per unit of land, without much concern for water use efficiency in economic terms, if the land is limited. This however does not mean that water use efficiency in economic terms would reduce, as empirical evidence is contrary to this. But, if water is limited, then farmers would be motivated to allocate more water to water-efficient crops, or crops that give higher returns per unit of water (Kumar, 2005; Lorite & Arriaza Balmón, 2008). Hence, in natural water-scarce regions of South Asia, pricing of water would bring about both physical efficiency and economic efficiency of water use.

A study that examined the decoupling of agricultural subsidies on irrigation water use and irrigation water productivity in Spain showed that with an increase in the price of irrigation water, the area under crops with high water requirements like cotton, sugar beet, and maize has been reduced almost by half, while the area under crops with low irrigation requirements such as winter cereals, sunflower or olive has increased by 37%. After the decoupling of the EU cotton subsidies in 2006, the cotton-growing practices have shifted toward a less water-intensive production system. For instance, the irrigation water release against the irrigation demand for cotton reduced from 0.80 to 0.50. The irrigation water productivity for cotton increased from around 0.7 to 1.0 € per m^3 in the irrigation season 2006/2007 (Lorite & Arriaza Balmón, 2008).

8.4.4.2 Energy pricing for groundwater pumping

Private well irrigation occupies large areas in India, Pakistan, Bangladesh, and Nepal Terai (Clauses & Molle, 2016). Electricity supplied in the farm sector for groundwater pumping is heavily subsidized in all these South Asian countries (except Pakistan). There is a heavy subsidy for diesel also in all these countries. Groundwater irrigation will not be possible in most of these regions without electricity, and electric pumps account for a major share of the well irrigation in India, due to deep water table conditions existing in semiarid and arid regions. While wells run by electric pumps account for a major share of the total well-irrigated area, electricity pricing is based on the connected load in most Indian states—barring West Bengal, and nearly 50% of the connections in Gujarat (Kumar et al., 2011). While electricity charges for groundwater pumping are very high in Pakistan, most farm wells are run by diesel pumps (Qureshi, McCornick, Sarwar, & Sharma, 2010). In Bangladesh, groundwater irrigation is dominated by shallow wells, and run mostly by diesel engines. Hence, one can expect water use efficiency to be good among well irrigators of Pakistan and Bangladesh. Nevertheless, greater efficiency in the use of energy and water could be achieved through the removal of subsidies in diesel.[i] In Sri Lanka, well irrigation accounts for a small fraction of the total irrigated area. Therefore, the only country where the introduction of metered power connection will make a remarkable impact on the efficiency of the use of water and electricity in near future is India.

Theoretically, pro-rata electricity pricing should bring about efficiency improvements in water use (Malik, 2008; Saleth, 1997), raising crop water productivity (kg per m^3). Empirical evidence from the field shows that it also raises water productivity in economic terms with the economic use of various inputs that minimize input costs. Also, the farmers who were confronted with a positive marginal cost of using water were found to be allocating more water to grow crops that produce income returns per unit of land (Rs. per ha), which also incidentally have higher water productivity in Rs. per m^3 of water. This means that there would be the same crop output in quantitative terms, and net income from a lesser amount of irrigation water. So, pro-rata pricing of electricity would be socioeconomically viable, provided electricity supply is of good quality and is reliable (Kumar et al., 2011).

---

[i] Empirical analysis shows that farmers who purchase water from diesel well owners, and who are confronted with higher marginal costs of using water, use water more efficiently than the pump owners and obtain higher water productivity in both physical and economic terms.

In water-scarce, but land rich regions, when water productivity improves due to pro-rata pricing, farmers might expand the area under cultivation. So, the same amount of water would provide far higher crop returns, in both quantitative terms and value terms (Kumar et al., 2011). A study in the US using modeling of production function showed a substantial reduction in the volume of groundwater pumped for irrigation with a rise in the price of electricity supplied (Pfeiffer & Lin, 2013). Hence, efficient pricing of electricity supplied for groundwater pumping promotes agricultural productivity growth.

8.4.5 Institutional changes

8.4.5.1 Water restrictions and water rights

Instruments such as the pricing of water and electricity alone may not ensure sustainability in water use, though it would bring about efficiency improvements in water use, as farmers can increase their water use to maximize their net returns (Kumar and Singh, 2001; Kumar, 2005). The impact of efficiency improvements on crop water productivity cannot be expected to be high. But, if water allocation to the fields is rationed, then farmers would be motivated to allocate more water to crops that are low water consumption, and simultaneously yield higher return per unit of water (Rs. per m^3) to maximize their income, because of the high opportunity cost of using water,[j] or sell it to their neighbors at a price decided by the market.

Water allocation to individual farmers' fields can be rationed through (i) restrictions in supply hours; and (ii) instituting "water rights or water entitlements," which are well defined in volumetric terms. The latter would require measurements at the field inlets. When water rights are defined and made tradable, the price at which water is traded would reflect the opportunity cost of using water (Frederick, 1993; Howe, Schurmeier, & Shaw, 1986). Tradable water rights, successfully introduced in Chile, Mexico, and California for both groundwater and surface water had improved water use efficiency (Rosegrant & Schleyer, 1996). The Chilean experience shows that it had benefited the small and marginal farmers, by allowing them to earn income through the sale of water to large farmers (Schleyer, 1996).

The most recent example of water rationing is of IGNP command in Rajasthan, which experienced a 50% cut in water delivery. But farmers adopted sprinkler irrigation in the command area with government subsidy,

[j] It is understood that with volumetric restrictions on water, the price for water would go up, which increases the opportunity cost of water usage.

and increased the area under cultivation, rather than reducing it (Amarasinghe et al., 2008). Similarly, in the Murray Darling Basin in Australia, following a 10-year-long drought (1999–2009), there has been a major reduction in the water supply to irrigators. But the introduction of tradable water rights encouraged the transfer of water from farms, which used irrigation water for low-value cereal crops, to farms that used it for high-value horticulture, raising the agricultural GDP of the basin by nearly $ 200 million.

Empirical research from different parts of the world has established the impact of deficit irrigation in enhancing crop water productivity (Deng, Shan, Zhang, & Turner, 2006; Geerts & Raes, 2009; Zwart & Bastiaanssen, 2004; Kumar & van Dam, 2013). When the water supply is restricted, the farmers would be motivated to limit their water application to a regime, which corresponds to maximum crop water productivity, rather than maximum yield, and therefore "deficit irrigation" can become a viable strategy in dryland regions.

In the case of electricity supply in the farm sector, this can be rationed along with the introduction of pro-rata pricing to bring about sustainable use of energy and groundwater (Malik, 2008). This can be analogous to the allocation of water rights. The impact will be similar to that of water rationing (Kumar et al., 2011; Zekri, 2008). The energy quota of individual farmers can be decided on the basis of sustainability considerations, based on the "safe yield" of the aquifers, the total area to be irrigated, and the landholding of individual farmers. Prepaid meters can be used to meter electricity consumption by farmers and can be activated through scratch cards or mobile SMS, or the internet to reduce the transaction cost of metering (Kumar et al., 2011). The use of prepaid electronic meters enables fixing of energy quota by the electricity utility (Zekri, 2008). Ultimately, its impact on crop production would depend on what we do with the saved water.

8.4.6 Improving water productivity in rainfed agriculture: Opportunities

As per official classification followed in most South Asian countries, most parts of the region are under rainfed agriculture. Afghanistan, Bhutan, and Sri Lanka are predominantly rainfed, with rainfed areas exceeding 80% of the total cropped area, as are India and Nepal (60%–70%). Irrigated agriculture is dominant in Pakistan (26%) and Bangladesh (45%). India, Bangladesh, Pakistan, Nepal, and Sri Lanka produce a wide range of agricultural and animal husbandry products. The forest area compared to land area is less than 30% in all countries except Bhutan where 86% of the land area is forest.

Of the South Asian countries, Pakistan and Afghanistan have the least forest cover (2.1%) (Rao et al., 2016).

With the rapid expansion in well irrigation, South Asian countries do not have purely "rain-fed areas" now in the strict sense of the term. But there are rain-fed crops in many regions, including central and peninsular India, eastern India, Bangladesh, and Sri Lanka. This is because some crops are always irrigated in every region, though some farmers might be growing those crops under the rain-fed condition there. Often, farmers who do not have irrigation facilities resort to the purchase of water to provide critical supplementary irrigation. An example is a cotton growing in Maharashtra and Madhya Pradesh. However, the situation with regard to the extent of irrigation keeps varying with rainfall variability. In a high rainfall year, certain crops might give high yields without irrigation, whereas in a low rainfall or drought year, securing optimum yield would not be possible without supplementary irrigation.

As shown by Droogers, Seckler, and Makin (2001), most of the areas which have low to moderate rain-fed yield potential are in central India and the Peninsular region. The potential for rain-fed farming is very low in almost entire Pakistan and Afghanistan. There are regions where the yield potential is high. Examples are the Bangladesh deltaic region, part of Nepal Tarai, western Ghat region (source: based on Droogers et al., 2001, figure 7, p. 11). Among these, the central Indian belt deserves special mention. This region is dominated by tribes, who are the first or second-generation agriculturists (Phansalkar & Verma, 2005). In spite of abundant natural resources, by and large, the population in this region is not able to improve their farming considerably, owing to their peculiar cultural and socioeconomic conditions. Instead, they mostly practice subsistence farming and grow most crops under rain-fed conditions. The development of water resources for irrigation is poor in these regions; the use of modern farming practices including the use of fertilizers and pesticides and crop technologies is extremely low. The result is that productivity is low for cereals, and total factor productivity growth is also very poor. The other food grain crops grown extensively in this region have low productivity (Amarasinghe & Sharma, 2009). Hence, this region is characterized by agricultural backwardness.

Analysis by Amarasinghe and Sharma shows that there are 208 districts where the average consumptive use of water for food grain production is low (below 300 mm), due to larger areas under rain-fed coarse grains like pulses such as green gram, black gram, and horse gram. These crops give very low grain yields, resulting in low WP. The study which used analysis of district-level aggregate data on crop outputs, and average consumptive use of water (CWU) also shows that supplementary irrigation can boost both yield and WP significantly (Amarasinghe & Sharma, 2009). This would be through farmers shifting from short-duration food grain crops to long-duration irrigated crops such as wheat in winter; and rainy season paddy to irrigated paddy.

This tribal region forms the upper catchment of important river basins in India such as the Mahanadi, Godavari, Tapi, Mahi, Narmada, Krishna, Sabarmati, and Banas, spread over the states of Orissa, Madhya Pradesh, Chattisgarh, Maharashtra, Gujarat, and Rajasthan. These regions form the rich catchments responsible for a major chunk of the basin yields, which are appropriated for downstream uses (CWC, 2017; Kumar et al., 2006). However, the flows in some of these basins are already exploited to their full potential for irrigation and other uses through storage and diversion systems, and further exploitation in the upstream areas would be at the cost of downstream uses. They are Mahi, Krishna, Sabarmati, and Banas (Kumar et al., 2006).

Large-scale irrigation projects are coming up in the Narmada river basin, where 29 large, 125 medium, and around 3000 minor schemes are being planned for construction. The work of some of them is already completed. The percentage cropped area currently irrigated in the basin is very small. The rain-fed crops occupy large areas in the basin (Ranade & Kumar, 2004). The rain-fed crops account for a major chunk of the basin's water economy. Many of these crops are those which need supplementary irrigation to realize the yield potential. The irrigated crops and crops receiving supplementary irrigation in the basin have much higher water productivity compared to rain-fed ones in both physical and economic terms. Once built, these irrigation schemes will be able to bring the rain-fed areas in the basin under irrigated production thereby raising crop water productivity. The productivity impacts would be visible on crops such as irrigated cotton, pulses such as gram, black gram, and green gram, and cereals such as paddy and wheat (Kumar, 2010).

There are still a few basins in South Asia, where small-scale water resources development is possible without any negative effects on the downstream. In India, such basins include the Tapi, Mahanadi, and a few small river basins (Karjan and Damanganga) in South Gujarat; the upper catchments of the west-flowing rivers of Western Ghats; the hilly and mountainous region of northeast and the north—which fall under the sub-Himalayan region; and the Godavari.

In Bangladesh, Chittagong Hill Tract, which receives more than 2500 mm rainfall and where farmers still practice *jhum* cultivation (Nath, Inoue, & Chakma, 2005), is suitable for small water harvesting for irrigating vegetables and fruits during the summer months. In Nepal, the entire mountainous region and the midlands are suitable for small water harvesting, where vegetables, fruit crops, and paddy can be provided with supplementary irrigation. Out of the areas mentioned above, in areas where the current rain-fed yields are less than the potential, small water harvesting systems can be used for supplementary irrigation of the rain-fed crops. Such areas include the central Indian belt, the hilly and mountainous regions of Nepal, and the hill regions of Bangladesh.

Since the geohydrological environment is generally not very favorable for the storage of the harnessed water underground due to hard rock strata in these hilly regions of the upper catchments, water can be stored in small-scale reservoirs such as anicuts, check dams, ponds, and tanks. But the water resources being harnessed should be put to beneficial uses immediately after harvest. The reason being that the potential evaporation rates are high in most parts of these regions even during the monsoon (Kumar et al., 2006) and a significant proportion of the stored water can be lost to evaporation. This means the best option is to use this harvested water for the supplementary irrigation of Kharif crops. This would increase evapotranspiration with a disproportionate increase in yield, leading to water productivity gain.

In peninsular India, rain-fed crops are still grown in many parts due to low and erratic rainfall and poor surface water availability, and groundwater endowment. This region is mostly underlain by hard rock formations. The problem of natural water scarcity in the basins is compounded by the demand for water far exceeding the renewable water resources. But, the only open basin where water resources development, including small water harvesting, could increase the utilizable water resources is the Godavari. Augmentation in water resources would be possible and the same water could be used to bring rain-fed crops under irrigation to boost crop yields. In any case, large-scale water resource development projects based on river lifting are coming up in this region and would help expand irrigated agriculture boosting crop water productivity.

In addition to the low water (ET) consuming short duration rain-fed (food) crops and rain-fed (food) crops that experience water stress, there are rain-fed crops that have a moderate consumptive use of water (300–425 mm) in 117 districts of India. These crops, which are essentially long duration, fine cereals, are concentrated in eastern and central India. After Amarasinghe and Sharma (2009), the yield gap of these food grain crops is very high. The use of better crop technologies and better inputs could also result in a significant improvement in water productivity through yield enhancement, which would be the effect of nutrients and proportion of the ET being used up for transpiration.

8.5 Enhancing agricultural water productivity in Sub-Saharan Africa

As we have discussed earlier, unlike south Asia, SS Africa has a very small proportion of its cultivable area under irrigation. As per Siebert et al. (2010), a total of 5.41 m per ha was irrigated in SS Africa, against 7.198 m per ha of the area actually equipped for irrigation.

Therefore, the strategies for improving crop water productivity in this region will have to be very different from that of South Asia. The major thrust has to be on increasing the level of utilization of the available water in the river basins to increase the irrigation potential, mainly in the semiarid Sahelian zone and also in the Savannah zone. Water resource development will continue to be a priority in the region for several decades to come. Once irrigation is assured, the farmers, who otherwise hesitate to take up crop cultivation due to fear of loss of their investments, will be encouraged to put more area under cultivation. With area expansion, more of the available green water and blue water could be beneficially used in crop production. This will improve the water use efficiency and water productivity at the basin level, as less amount of water from the river basins of the region will be lost to the natural sinks.

Of the total irrigation of 5.51 m per ha, groundwater irrigation is only 0.34 m per ha (i.e., only 6%). The rest is gravity-based systems (Siebert et al., 2010). If the public investment in irrigation goes for large and medium gravity-based systems, the opportunities to improve the productivity of water use may not be very high due to the inherent difficulties in establishing control over water application in systems. It is quite well established that the use of efficient irrigation technologies such as drips and sprinklers largely pick when water is available at the farm inlets under pressure. But supplying water under pressure in surface irrigation systems will be far more expensive than supplying it under gravity (Renault, Wahaj, & Smits, 2013). Therefore, it is important that the farmers in and around gravity irrigation commands are encouraged to develop groundwater resources through proper economic incentives including subsidized electricity (but supplied on a pro-rata basis through metering) or diesel. Namara, Gebregziabher, Giordano, and De

Fraiture (2013) note that private smallholder irrigation in SS Africa is practiced mainly by the wealthier farmers and that the development of groundwater irrigation requires targeted and deliberate public-policy interventions and institutional support focusing on the more marginal farmers. Since gravity irrigation would result in an improved recharge to shallow groundwater through seepage from canals and irrigation return flows (source: based on Watt, 2008; Renault et al., 2013), the investment risks associated with groundwater development arising from the uncertainty about resource availability will be least.

Access to well irrigation will make the adoption of pressurized irrigation systems such as drips, sprinklers, and the Californian system feasible. On the other hand, since the cost of abstracting water from deep aquifers is high in the region due to the high cost of well-drilling (Foster, Tuinhof, & Garduño, 2006) and the high cost of electricity/diesel required for pumping water, it becomes imperative to use efficient irrigation systems to improve the efficiency of water application, thereby reducing the irrigation costs. Hence while inefficient use of water through gravity irrigation system will promote well irrigation which can be made very efficient through the use of technologies. It is also essential that high-value crops that are also amenable to drips and sprinklers are promoted in the well-irrigated areas to raise the return on investments.

Field trials carried out in six Sub-Saharan African countries (viz., Burkina Faso, Mali, Senegal, The Gambia, Niger, and Mauritania) involving the comparison of traditional bucket method of irrigation with efficient irrigation systems (drip systems and Californian method) for vegetable crops showed much higher water productivity (in both physical and economic terms) in crop production under drips and Californian system, compared with the traditional method. The results of the field trials conducted comparing the Californian system with bucket method, and drip system with bucket method, are presented in Tables 8.2 and 8.3, respectively. Water productivity ($ per m^3) under the Californian system was two to three times higher than that of the bucket method (Table 8.2), and that under drip method was 1.5 to 2 times higher than that of the bucket method (Table 8.3) (source: Qureshi & Ismail, 2016).

Since the cost of production and supply of irrigation water through canals and wells is generally very high in the region (Namara et al., 2013), it is important to provide capital subsidies for efficient irrigation technologies (especially drip irrigation) that are very expensive, particularly in the semiarid Sahelian region. However, the extent of capital subsidy

Table 8.2 Benefits of California method of irrigation over Bucket method of irrigation.

| Attributes | Benefit of Californian system over Bucket method | | | |
	Burkina Faso (Onion)	Mali (Onion)	Mauritania (Tomato)	Senegal (Onion)
Reduction in irrigation water applied (m³/ha)	1727	2140	1600	2680
Improvement in crop yield (kg/ha)	475	6380	18,700	5200
Rise in physical water productivity (kg/m³)	0.23	2.91	5.28	1.26
Rise in economic water productivity ($/m³)	0.11	1.46	2.66	0.74

Source: Based on Qureshi, A.S. & Ismail, S. (2016) Improving agricultural productivity by promoting low-cost irrigation technologies in Sub-Saharan Africa. Global Advanced Research Journal of Agricultural Science, 5 (7), 283–292.

Table 8.3 Benefits of drip method of irrigation over Bucket methods of irrigation.

| Attributes | Benefit of Californian system over Bucket method | | | |
	Burkina Faso (Tomato)	The Gambia (Tomato)	Niger (Onion)	Mail (Tomato)
Reduction in irrigation water applied (m³/ha)	1450	800	620	1620
Improvement in crop yield (kg/ha)	1870	5735	600	18,507
Rise in physical water productivity (kg/m³)	1.72	0.94	0.12	3.92
Rise in economic water productivity ($/m³)	0.85	0.47	0.02	1.96

Source: Based on Qureshi, A. S. & Ismail, S. (2016) Improving agricultural productivity by promoting low-cost irrigation technologies in Sub-Saharan Africa. Global Advanced Research Journal of Agricultural Science, 5 (7), 283–292.

provided for the system (per unit area) should be decided on the basis of the economic value of the externality induced by the technology on the society through irrigation water saving, and in such a way the private cost of the system becomes less than the incremental economic benefit derived from the use of the system.

The real water saving from the use of such technologies would come from a reduction in nonbeneficial evaporation, and nonrecoverable deep percolation. The extent of real water-saving would depend mainly on the crop, climate, geohydrology (depth to the water table), and irrigation technology used. In arid areas with a deep water table, there could be significant real water saving through drip systems (Kumar, 2016; Kumar & van Dam, 2013).

Small water harvesting systems can also be promoted in regions that experience moderately high rainfall with low interannual variability (in the Savannah), in situations where the harvested water can be diverted for providing supplementary irrigation to crops (Rockström, Barron, & Fox, 2002). However, the river basins where such systems are promoted should be "open" in the sense that the basin has a sufficient amount of untapped runoff (or uncommitted flows) that can be harvested without causing adverse consequences for the existing water impoundment and diversion systems downstream. As studies in India have shown, the lower the interannual variability in rainfall and runoff and degree of aridity in such basins, the better will be the economic viability of such small water harvesting systems. It is also important to note that the cost of water harvesting/development would increase as more and more water from the basin gets appropriated (Kumar et al., 2006; Kumar, Patel, et al., 2008). The investment decisions for small water harvesting systems should be based on the criterion that the net surplus value product from the use of water for irrigation and other high value uses ($ per m^3 of water consumed) is higher than the unit cost of harvesting ($ per m^3) (Kumar & van Dam, 2013). Small water harvesting systems can then be combined with microirrigation systems to irrigate high-value crops in the neighborhood to promote physically and economically efficient uses of the expensive water.

8.6 Conclusions

In this chapter, we have dealt with two distinct regions, viz., South Asia and Sub-Saharan Africa, to explore the opportunities for improving water productivity in agriculture. When we deal with agricultural production and productivity, the South Asian region is not a monolithic entity. The region is marked by several agroecologies. The relative availability of water in relation to land also changes from region to region. There are semiarid and arid low rainfall regions with plenty of land resources that are suitable for cultivation, but internal renewable water resources are very scarce. On the other extreme, there are regions where water is available in plenty, and land resources are very scarce, and the limited land resources also face ecological constraints such as flooding and poor sunshine for several months of the year.

Irrigated agriculture is practiced in both low rainfall semiarid (and arid), water-scarce regions, and high-rainfall, subhumid water-rich regions. The need for improving water productivity is obviously greater in the former.

There are several options for raising water productivity in irrigated agriculture in South Asia. They fall under the following broad categories: (1) technological change for improving the technical efficiency improvements, which include the development of varieties with high genetic yield potential, development of varieties that are resistant to droughts, floods, waterlogging, and salinity; introduction of efficient irrigation technologies; improved crop and farm management practices for reducing yield gaps for major cereals such as wheat and rice, fertigation for improved nutrient use efficiency; (2) other measures that include crop management, crop shifts, taking advantage of the climate to decide on the cropping patterns for different regions, water delivery control, improving the quality and reliability of irrigation; (3) policy changes for affecting improvements in technical efficiency and allocative efficiency—irrigation water pricing and electricity pricing for groundwater pumping; and (4) institutional changes—restrictions on water allocation and water rights.

In several Indian states, Pakistan and Sri Lanka, irrigation water are highly subsidized. So is the electricity supplied for groundwater pumping for irrigation. Water rights systems or the practice of restricting volumetric water allocation for irrigation do not exist anywhere in South Asia. Because of these reasons, the farmers of these countries have no special incentive to use water efficiently, unless there are physical constraints to overusing water due to water shortages. The opportunities for affecting technology changes for improving the physical efficiency of water use in agriculture through institutional and policy measures are considerably high.

As regards rainfed agriculture, there are large regions in India covering about 228 districts where short-duration food crops are grown under rainfed conditions due to a lack of irrigation facilities, resulting in low yields. With the provision of supplementary irrigation, a shift to long-duration crops (rice and wheat) would be possible, increasing both yield and water productivity considerably. With the provision of irrigation facilities, a shift from rainfed paddy to irrigated paddy is possible in large areas. With large-scale water resources development happening in Central and South-Central India, significant areas under rainfed crops could be brought under irrigated production, raising both yield and water productivity.

In addition to the short-duration rain-fed (food) crops, there are rain-fed crops that have moderate consumptive use of water (300–425 mm) in 117 districts of India. These crops, which are essentially long duration, fine cereals, are concentrated in eastern India and central India. They show very high yield gaps. The use of better crop technologies and better inputs could also result in a significant improvement in water productivity through yield enhancement, which would be the effect of nutrients and proportion of the ET being used up for transpiration. In the case of eastern India, which is water-rich, rather than raising water productivity, raising the yield of these cereal crops is crucial for taking the region out of agricultural stagnation and tackling high rural poverty. Boost in agricultural production in eastern India can ease the pressure on the water resources of agriculturally prosperous water-scarce regions of north-western, western, and southern India.

In Sub-Saharan Africa, the interventions to promote water productivity improvements should focus on the following: (i) increasing the utilization of available surface water resources for irrigation development in the semiarid (Sahelian) zone and the Savannah; (ii) promoting well irrigation in the command areas of large and medium gravity irrigation systems through economic incentives; (iii) small water harvesting systems in the zones where rainfall is reliable and aridity is less, as a source for supplementary irrigation to high-value crops grown in the closest fields, combined with microirrigation systems (especially drips); and (iv) promoting microirrigation in well commands to irrigate high-value crops to raise the physical and economic efficiency of water use.

An appropriate subsidy structure for promoting microirrigation system shall be designed so that the capital cost of the system is less than the value of the incremental benefits accrued by the farmers from the use of the system. Hence it is obvious that in water-scarce regions, the capital subsidy for efficient irrigation systems such as drips should be kept high in view of the high social benefits accrued through water-saving, and the private cost of the system becomes less than the incremental economic benefit derived from the use of the technology.

References

Aggarwal, P. K., Hebbar, K. B., Venugopalan, M. V., Rani, S., Bala, A., Biswal, A., & Wani, S. P. (2008). *Quantification of yield gaps in rain-fed rice, wheat, cotton and mustard in India*: Global Theme on Agroecosystems Report no. 43.

Aggarwal, P. K., Kalra, N., Bandyopadhyay, S. K., & Selvarajan, S. (1995). A systems approach to analyze production options for wheat in India. In J. Bouma, et al. (Eds.), *Systems approach for sustainable agricultural development*. The Netherlands: Kluwer Academic Publishers.

Allen, R. G., Willardson, L. S., & Frederiksen, H. (1997). Water use definitions and their use for assessing the impacts of water conservation. In J. M. de Jager, L. P. Vermes, & R. Rageb (Eds.), *Sustainable irrigation in areas of water scarcity and drought* (pp. 72–82). England: Oxford. September 11–12.

Amarasinghe, U., Bhaduri, A., Singh, O. P., Ojha, A., & Anand, B. K. (2008). Cost and benefits of intermediate water storage structures: Case study of Diggies in Rajasthan. In M. D. Kumar (Ed.), *Vol. 1. Managing water in the face of growing scarcity, inequity and declining returns: Exploring fresh approaches. Proceedings of the 7th annual partners meet, IWMI TATA water policy research program, ICRISAT, Patancheru, Hyderabad, India* (pp. 51–66). 2–4 April 2008.

Amarasinghe, U. A., & Sharma, B. R. (2009). Water productivity of food grains in India: Exploring potential improvements. In M. D. Kumar, & U. A. Amarasinghe (Eds.), *Strategic analyses of the National River Linking Project (NRLP) of India series 5. Water productivity improvements in Indian agriculture: Potentials, constraints and prospects* (p. 163).

Bhaskara Rao, M. (2011). Drought scenario in South Asia–an overview of integrated water resources development & management techniques for drought management in South Asia. In P. K. Mishra, O. M. Satendra, & B. Venkateswarlu (Eds.), *Techniques of water conservation & rainwater harvesting for drought management* (p. 714). SAARC Training Program. July 18, e29.

Bruinsma, J. (2009, June). The resource outlook to 2050: By how much do land, water and crop yields need to increase by 2050. In *Vol. 2050. Expert meeting on how to feed the world in* (pp. 24–26).

Byerlee, D., & Traxler, G. (1996). The role of technology spill overs and economies of size in the efficient design of agricultural research systems. Agricultural science policy: Changing global agendas.

Central Water Commission. (2017). *Reassessment of water availability in India using space inputs.* New Delhi: Basin Planning and Management Organisation. Central Water Commission. October 2017.

Clauses, A., & Molle, F. (2016). *Groundwater governance in Asia and the Pacific, IWMI project report 4, groundwater governance in the Arab world.* IWMI and USAID. December.

Deng, X. P., Shan, L., Zhang, H., & Turner, N. C. (2006). Improving agricultural water use efficiency in arid and semiarid areas of China. *Agricultural Water Management, 80*(1–3), 23–40.

Droogers, P., Seckler, D., & Makin, I. (2001). *Estimating the potential of rain-fed agriculture, working paper # 20.* Colombo, Sri Lanka: International Water Management Institute.

Evenson, R. E., Pray, C. E., & Rosegrant, M. W. (1999). *Agricultural research and productivity growth in India.* Research Report 109, International Food Policy Research Institute.

Fan, Y., Wang, C., & Nan, Z. (2018). Determining water use efficiency of wheat and cotton: A meta-regression analysis. *Agricultural Water Management, 199*, 48–60. https://doi.org/10.1016/j.agwat.2017.12.006.

Food and Agriculture Organization of the United Nations. (1986). Irrigation in Africa south of the Sahara, FAO Investment Centre Technical Paper 5, Food and Agriculture Organization, Rome, Italy.

Foster, S., Tuinhof, A., & Garduño, H. (2006). *Groundwater development in Sub-Saharan Africa: Strategic overview of key issues and major needs. Case profile collection 15.* Washington, D.C., USA: The World Bank.

Frederick, K. D. (1993). Balancing water demands with supplies: The role of management in a world of increasing scarcity. In *World Bank technical paper (No. PB-93-180750/XAB; WORLD-BANK-TP—189).* Washington, DC (United States): International Bank for Reconstruction and Development.

Geerts, S., & Raes, D. (2009). Deficit irrigation as an on-farm strategy to maximize crop water productivity in dry areas. *Agricultural Water Management, 96*(9), 1275–1284.

Government of Pakistan (GOP). (2012). *Canal water pricing for irrigation in Pakistan: Assessment, issues and options*. Planning Commission, Govt. of Pakistan, Islamabad.

Haisman, B. (2005). Impacts of water rights reform in Australia. In B. R. Bruns, C. Ringler, & R. S. Meinzen-Dick (Eds.), *Water rights reform: Lessons for institutional design* International Food Policy Research Institute.

Howe, C. W., Schurmeier, D. R., & Shaw, W. D., Jr. (1986). Innovative approaches to water allocation: The potential for water markets. *Water Resources Research, 22*(4), 439–445.

Hussain, I., Sakthivadivel, R., Amarasinghe, U., Mudasser, M., & Molden, D. (2003). *Land and water productivity of wheat in the western Indo-Gangetic plains of India and Pakistan: A comparative analysis, research report # 65*. Colombo, Sri Lanka: International Water Management Institute.

Jin, M., Zhang, R., Sun, L., & Gao, Y. (1999). Temporal and spatial soil water management: A case study in the Heilonggang region, PR China. *Agricultural Water Management, 42*, 173–187.

Kumar, M. D. (2005). Impact of electricity prices and volumetric water allocation on energy and groundwater demand management: Analysis from Western India. *Energy Policy, 33*(1), 39–51.

Kumar, M. D. (2010). *Managing water in river basins: Hydrology, economics, and institutions*. Oxford University Press.

Kumar, M. D. (2016). Water saving and yield enhancing micro irrigation technologies in India: Theory and practice. In P. K. Viswanathan, M. D. Kumar, & A. Narayanamoorthy (Eds.), *Springer series in business and economics. Micro irrigation Systems in India: Emergence, status and impacts* (pp. 13–36). Singapore: Springer Singapore.

Kumar, M. D., Ghosh, S., Patel, A., Singh, O. P., & Ravindranath, R. (2006). Rainwater harvesting in India: Some critical issues for basin planning and research. *Land Use and Water Resources Research, 6*. 1732-2016-140267.

Kumar, M. D., Patel, A., Ravindranath, R., & Singh, O. P. (2008). Chasing a mirage: Water harvesting and artificial recharge in naturally water-scarce regions. *Economic and Political Weekly, 43*(35), 61–71.

Kumar, M. D., Scott, C. A., & Singh, O. P. (2011). Inducing the shift from flat-rate or free agricultural power to metered supply: Implications for groundwater depletion and power sector viability in India. *Journal of Hydrology, 409*(1–2), 382–394.

Kumar, M. D., & Singh, O. P. (2001). Market instruments for demand management in the face of growing scarcity and overuse of water in India. *Water Policy, 5*(3), 86–102.

Kumar, M. D., Singh, O. P., Samad, M., Purohit, C., & Didyala, M. S. (2008). Water productivity of irrigated agriculture in India: Potential areas for improvement. In *Proceedings of the 7th annual partners' meet of IWMI-Tata water policy research program, international water management institute, South Asia regional office, ICRISAT campus, Hyderabad*. 2-4 April 2008.

Kumar, M. D., Trivedi, K., & Singh, O. P. (2009). Analyzing the impact of quality and reliability of irrigation water on crop water productivity using an irrigation quality index. In M. D. Kumar, & U. Amarasinghe (Eds.), *Strategic analyses of the National River Linking Project (NRLP) of India series 5. Water productivity improvements in Indian agriculture: Potentials, constraints and prospects* (p. 163).

Kumar, M. D., Turral, H., Sharma, B. R., Amarasinghe, U., & Singh, O. P. (2008). Water saving and yield enhancing micro-irrigation technologies in India. When and where can they become best bet technologies? In *Proceedings of the 7th annual partners' meet of IWMI-Tata water policy research program. International water management institute, South Asia regional office, ICRISAT campus, Hyderabad*. 2-4 April 2008.

Kumar, M. D., & van Dam, J. C. (2009). Improving water productivity in agriculture in India: Beyond 'more crop per drop'. In M. D. Kumar, & U. Amarasinghe (Eds.), *Strategic analyses of the National River Linking Project (NRLP) of India series 5. Water productivity improvements in Indian agriculture: Potentials, constraints and prospects* (p. 163).

Kumar, M. D., & van Dam, J. C. (2013). Drivers of change in agricultural water productivity and its improvement at basin scale in developing economies. *Water International, 38*(3), 312–325.

Kumar, M. D., Vedantam, N., Puri, S., & Bassi, N. (2012). *Irrigation efficiencies and water productivity in sugarcane in Godavari River basin, Maharashtra, final report submitted to the world wild Fund for Nature.* Hyderabad: Institute for Resource Analysis and Policy.

Kumar, M. D., Viswanathan, P. K., & Bassi, N. (2014). Water scarcity and pollution in South and Southeast Asia: Problems and challenges (Chapter 12). In P. G. Harris & G. Lang (Eds.), *Routledge Handbook of Environment and Society in Asia* (pp. 197–215). London: Routledge, Taylor & Francis Group.

Lal, R. (2005). Climate change, soil carbon dynamics, and global food security. In R. Lal, B. Stewart, N. Uphoff, et al. (Eds.), *Climate change and global food security* (pp. 113–143). Boca Raton, FL: CRC Press.

Lankford, B. (2012). Fictions, fractions, factorials and fractures; on the framing of irrigation efficiency. *Agricultural Water Management, 108* (2012), 27–38.

Loomis, R. S., & Connor, D. J. (1996). *Productivity and management in agricultural systems.* Cambridge, New York and Australia: Cambridge University Press.

Lorite, I. J., & Arriaza Balmón, M. (2008). *Effects of the decoupling of the subsidies on agricultural water productivity.* (No. 725-2016-49588).

Malik, R. P. S. (2008). Energy regulations as a demand management option: Potentials, problems, and prospects. In R. M. Saleth (Ed.), *Strategic Analyses of the National River Linking Project (NRLP) of India, Series 3. Promoting irrigation demand management in India: Potentials, problems and prospects. Strategic analyses of the National River Linking Project (NRLP) of India.* Colombo, Sri Lanka International Water Management Institute (IWMI).

Namara, R. E., Gebregziabher, G., Giordano, M., & De Fraiture, C. (2013). Small pumps and poor farmers in sub-Saharan Africa: An assessment of current extent of use and poverty outreach. *Water International, 38*(6), 827–839.

Nath, T. K., Inoue, M., & Chakma, S. (2005). Shifting cultivation (jhum) in the Chittagong Hill tracts, Bangladesh: Examining its sustainability, rural livelihood and policy implications. *International Journal of Agricultural Sustainability, 3*(2), 130–142.

Palanisami, K., Gemma, M., & Ranganathan, C. R. (2008). Stabilisation value of groundwater in tank irrigation systems. *Indian Journal of Agricultural Economics, 63*(1), 126–134. 902-2016-67957.

Pathak, H., Ladha, J. K., Aggarwal, P. K., Peng, S., Das, S., Singh, Y., ... Aggarwal, H. P. (2003). Trends of climatic potential and on-farm yields of rice and wheat in the Indo-Gangetic Plains. *Field Crops Research, 80*(3), 223–234.

Perry, C. (2001). Water at any price? Issues and options in charging for irrigation water 1. *Irrigation and Drainage: The Journal of the International Commission on Irrigation and Drainage, 50*(1), 1–7.

Pfeiffer, L., & Lin, C. (2013). *The effects of energy prices on groundwater extraction in agriculture in the High Plains aquifer.* No. 328-2016-12730.

Phansalkar, S. J., & Verma, S. (2005). *Mainstreaming the margins: Water-centric livelihood strategies for revitalizing tribal agriculture in Central India.* Angus and Grapher.

Pingali, P., & Heisey, P. W. (1999). Cereal crop productivity in developing countries: Past trends and future prospects (No. 7682). CIMMYT: International Maize and Wheat Improvement Center.

Qureshi, A. S. (2002). Water resources management in Afghanistan: The issues and options, Research Report # 49. International Water Management Institute, Colombo, Sri Lanka.

Qureshi, A. S., & Ismail, S. (2016). Improving agricultural productivity by promoting low-cost irrigation technologies in Sub-Saharan Africa. *Global Advanced Research Journal of Agricultural Science, 5*(7), 283–292.

Qureshi, A. S., McCornick, P. G., Sarwar, A., & Sharma, B. R. (2010). Challenges and prospects of sustainable groundwater management in the Indus Basin, Pakistan. *Water Resources Management, 24*(8), 1551–1569.

Ranade, R., & Kumar, M. D. (2004). Narmada water for groundwater recharge in North Gujarat: Conjunctive management in large irrigation projects. *Economic and Political Weekly, 39*(31), 3510–3513.

Rao, C. S., Gopinath, K. A., Rao, C. R., Raju, B. M. K., Rejani, R., Venkatesh, G., & Kumari, V. V. (2016). Dryland agriculture in South Asia: Experiences, challenges and opportunities. In M. Farooq, & K. H. Siddique (Eds.), *Innovations in dryland agriculture* Springer.

Renault, D., Wahaj, R., & Smits, S. (2013). *Multiple uses of water services in large irrigation systems: Auditing and planning modernization.* The MASSMUS Approach: Food and Agriculture Organization of the United Nations, Rome.

Rockström, J., Barron, J., & Fox, P. (2002). (2002) rainwater management for improving productivity among small holder farmers in drought prone environments. *Physics and Chemistry of the Earth, 27*, 949–959.

Rosegrant, M. W., & Schleyer, R. G. (1996). Establishing tradable water rights: Implementation of the Mexican water law. *Irrigation and Drainage Systems, 10*(3), 263–279.

Ruthenberg, H. (1980). *Farming systems of the tropics.* Oxford: Clarendon Press.

Saleth, R. M. (1997). Power tariff policy for groundwater regulation: Efficiency, equity and sustainability. *Artha Vijnana, 39*, 312–322.

Schleyer, R. G. (1996). Chilean water policy: The role of water rights, institutions and markets. *International Journal of Water Resources Development, 12*(1), 33–48.

Sharda, V. N. (2011). Soil and water conservation measures for checking land degradation in India. In P. K. Mishra, M. Osman, Satendra, & B. Venkateswarlu (Eds.), *Techniques of water conservation & rainwater harvesting for drought.*

Siebert, S., Burke, J., Faures, J. M., Frenken, K., Hoogeveen, J., Döll, P., & Portmann, F. T. (2010). Groundwater use for irrigation—A global inventory. *Hydrology and Earth System Sciences, 14*(10), 1863–1880.

Sikka, A. K. (2009). Water productivity of different agricultural systems. In M. D. Kumar, & U. Amarasinghe (Eds.), *Strategic analyses of the National River Linking Project (NRLP) of India series 5. Water productivity improvements in Indian agriculture: Potentials, constraints and prospects* (p. 163).

Sinclair, T. (1998). Options for sustaining and increasing the limiting yield-plateaus of grain crops. In *Paper presented at the NAS colloquium "plants and population: Is there time?", Irvine, CA, USA.* 5-6 December 1998.

Singh, V. P., Singh, P. K., & BHATT, L. (2014). Use of plastic mulch for enhancing water productivity of off-season vegetables in terraced land in Chamoli district of Uttarakhand, India. In *Vol. 13. Land capability classification and land resources planning using remote sensing and GIS 3* (pp. 68–72). 1.

Singh, R., van Dam, J. C., & Feddes, R. A. (2006). Water productivity analysis of irrigated crops in Sirsa district, India. *Agricultural Water Management, 82*(3), 253–278.

Thobanl, M. (1997). Formal water markets: Why, when, and how to introduce tradable water rights. *The World Bank Research Observer, 12*(2), 161–179.

Ullah, M. K., Habib, Z., Muhammad, S. (2001). *Spatial distribution of reference and potential evapotranspiration across the Indus Basin Irrigation Systems.* Working Paper 24. Lahore, Pakistan: International Water Management Institute.

Watt, J. (2008). *The effect of irrigation on surface-ground water interactions: Quantifying time dependent spatial dynamics in irrigation systems* (Doctoral dissertation). Charles Sturt University.

Xie, Z. K., Wang, Y. J., & Li, F. M. (2005). Effect of plastic mulching on soil water use and spring wheat yield in arid region of Northwest China. *Agricultural Water Management, 75*(1), 71–83.

Zekri, S. (2008). Using economic incentives and regulations to reduce seawater intrusion in the Batinah coastal area of Oman. *Agricultural Water Management, 95*(3), 243–252.

Zhang, P., Wei, T., Han, Q., Ren, X., & Jia, Z. (2020). Effects of different film mulching methods on soil water productivity and maize yield in a semiarid area of China. *Agricultural Water Management, 241*, 106382. https://doi.org/10.1016/j.agwat.2020.106382.

Zhang, F., Zhang, W., Qi, J., & Li, F. M. (2018). A regional evaluation of plastic film mulching for improving crop yields on the loess plateau of China. *Agricultural and Forest Meteorology, 248*, 458–468.

Zwart, S. J., & Bastiaanssen, W. G. (2004). Review of measured crop water productivity values for irrigated wheat, rice, cotton and maize. *Agricultural Water Management, 69*(2), 115–133.

CHAPTER 9

Constraints of improving crop water productivity in rainfed and irrigated systems in South Asia and Sub-Saharan Africa

9.1 Introduction

South Asia region has large areas that come under intensive irrigation by virtue of the presence of well-developed irrigation systems with canal networks and wells and tube wells. Out of the total of 92.419 m per ha of irrigated area, well irrigation covers an area of 50.593 m per ha, surface irrigation in 33.851 m per ha, and other sources in 7.975 m per ha (FAO, 2011). To name a few are the Indus basin irrigation system of Pakistan, supported by shallow and deep tube wells tapping the alluvium; Bhakra–Nagla irrigation system serving large areas in Punjab (India) and Haryana, supported by shallow and deep tube wells again tapping the alluvium; and gravity irrigation system in coastal Andhra Pradesh and Cauvery delta. These regions are also known for mono-cropping, high cropping intensity, and high agricultural productivity in terms of yield.

The South Asia region is also marked by large areas under rainfed conditions in most countries, barring Pakistan and Bangladesh (Rao et al., 2016) which include two distinct areas: high rainfall areas having high humidity with low inter-annual variability where crops do not require irrigation to mature in the major growing-season (known as Kharif in India, Maha season in Sri Lanka), and very low to medium rainfall areas having high aridity, with significant interannual variability, where crops often experience moisture stress due to insufficient rains or due to long dry spells between rainy days even during the main crop-growing season, in the absence of dependable irrigation sources. Large areas of West Bengal, Assam, Eastern UP, and Bihar in India, the wet zone of Sri Lanka, and large parts of Bangladesh fall under

Current Directions in Water Scarcity Research, Volume 3
https://doi.org/10.1016/B978-0-323-91277-8.00004-6

the first category. Low to medium yields mark these areas, depending on the agroecology. Large parts of central and western India and southern India, and vast areas of Afghanistan and Pakistan fall under the second category. These areas are marked by low crop yields.

It is widely believed that there is the uncontrolled use of irrigation water combined with the excessive use of fertilizers in these intensively irrigated areas, resulting in waterlogging and salinity, soil degradation and crop yields (that are lower than optimum) and very low water use efficiency and water productivity of crops (FAO, 2015; Yaduvanshi, Setter, Sharma, Singh, & Kulshreshtha, 2012). It is argued that water use efficiency and water productivity of crops grown in the areas could be easily improved through better farm water management, including limiting irrigation water delivery in certain cases, and improvement in soil quality and drainage.

Similarly, it is argued that poor watershed management and lack of focus on conservation of soils, in situ rainwater, and crops result in low yields and high crop failures in rainfed areas. It is further argued that crop water productivity, as well as crop yields in these areas, could be easily increased by rainwater harvesting, in situ water harvesting and soil and water conservation, evaporation prevention that increase the water availability at the plant level, and integrated soil, water, and crop management that increase the water intake capacity of the plants and that it is technically feasible and economically viable (Rockström et al., 2010). Such propositions for rainfed areas failed to make the essential distinction between the areas receiving high rainfall and areas receiving insufficient rainfall for crops to mature (Kumar, Reddy, Narayanamoorthy, Bassi, & James, 2018).

In the context of SS Africa, which has a very small fraction of its cultivable area under irrigation, many of the above arguments for improving productivity in rainfed areas such as watershed management largely echo in the academic and policy circles. The discussions on reducing yield gaps and improving water use efficiency have also focused on water harvesting for supplementary irrigation to rainfed crops (Rockström, Barron, & Fox, 2002).

In any case, there was little analysis of the constraints to improving agricultural water productivity in these regions. This chapter focuses on the constraints to improving water productivity in irrigated and rainfed areas without reducing economic benefits. The types of constraints included socioeconomic and financial and social in the case of rainfed areas, and physical, institutional, policy-related, market-related, technological, and environmental in the case of irrigated areas.

9.2 Constraints in the rain-fed agriculture

9.2.1 Socioeconomic and financial constraints

In the context of Burkina Faso and Kenya, Rockström et al. (2002) show that supplementary irrigation through water harvesting will have a remarkable effect on the productivity of water (expressed in kg/ET) for crops such as sorghum and maize. However, the research did not evaluate the incremental economic returns due to supplementary irrigation against the incremental costs of water harvesting. It also does not quantify the real hydrological opportunities available for water harvesting at the farm level and its reliability. The work by Scott, Silva-Ochoa, Florencio-Cruz, and Wester (2001) in the Lerma-Chapala basin in Mexico showed higher gross value product from crop production in areas with better allocation of water from water harvesting irrigation systems. But, their figures of surplus-value product which takes into account the cost of irrigation are not available from their analysis. In arid and semiarid regions, the hydrological and economic opportunities of water harvesting are often over-played. Recent work in India has shown that the cost of water harvesting systems would be enormous, and the reliability of supplies from it very poor in arid and semiarid regions of India, which are characterized by low mean annual rainfalls, very few rainy days, high inter-annual variability in rainfall and rainy days, and high potential evaporation leading to much higher variability in a runoff between good rainfall years and poor rainfall years (Kumar, Ghosh, Patel, Singh, & Ravindranath, 2006; Kumar, Patel, Ravindranath, & Singh, 2008).

As incremental returns due to yield benefits may not exceed the cost of the system, as indicated, the comparison between the unit cost of water harvesting and recharging schemes and the net returns from unit volume of water obtained in irrigated crops (Kumar, Patel et al., 2008), small and marginal farmers will not have the incentive to go for it. But, even if the benefits due to supplementary irrigation from water harvesting exceed the costs, it will not result in a basin-level gain in WP in economic terms in closed basins. The exception is when the incremental returns are disproportionately higher than the increase in ET. This is because, in a closed basin, an increase in beneficial ET at the place of water harvesting will eventually reduce the beneficial ET d/s, causing income losses there. Also, as noted by Kumar and van Dam (2013), incremental net benefit considerations can drive water harvesting at the basin scale only if there is no opportunity cost of harvesting.

In open basins, water harvesting and recharge schemes could be attempted to improve the water productivity of crops. But the following are prerequisites: (1) the harvested water is put to high valued use, making the system economically viable from the point of view of economic costs and the incremental benefits; or (2) the system is used to produce crops that provide very high social returns, especially in improving the regional food security and employment. In closed basins, it would be difficult to justify investments in water harvesting and recharge schemes from an economic perspective, unless the incremental returns due to the upstream interventions are far higher than the opportunity costs of downstream economic losses, and mechanisms are in place to compensate for these losses.

Unfortunately, the regions in South Asia, which are water-rich, have a very high concentration of native tribal population, be it in India or Nepal, or Bhutan or Bangladesh. They are used to growing subsistence crops like paddy and maize which have low economic returns and water productivity (Rs/m^3). Hence, most of the precondition for achieving water productivity gain through supplementary irrigation is not likely to be satisfied. This poses socioeconomic constraints. The situation in the humid zone and Savannah region of Sub-Saharan Africa, where farmers grow subsistence crops under rainfed conditions, are also similar, with the region ranking very low (HDI = 0.537) on human development indicators (source: http://hdr.undp.org/sites/default/files/Country-Profiles/MLI.pdf).

Investments needed for water harvesting and groundwater recharge schemes that can help improve water productivity in rain-fed farming systems are very high in terms of cost per cubic meter of water, even if they are economically viable or are able to generate high social returns. The poor tribes are least likely to mobilize these resources. Hence, they are financial constraints too. Large-scale government financing of water harvesting and groundwater recharge systems would therefore be required.

9.2.2 Social constraints

A disproportionately higher percentage of the rural proportion in South Asian countries lives in the rain-fed area—be it India, Bangladesh, or Nepal. These rain-fed areas include dryland areas where irrigation water is in short supply (like in Pakistan, western and south Indian peninsula) and areas where crops are grown under rain-fed conditions either because of lack of finance to invest in irrigation facilities (like in Bihar) or due to high rainfall (Kerala).

A large populations of Sub-Saharan Africa also live in rainfed areas, growing cereal crops (maize, sorghum, finger millet, and pearl millet, wheat, rice, and teff) with varying lengths of the growing period, depending on the amount of rainfall and the relative humidity (Macauley, 2015).

Water productivity improvement in rain-fed farming really matters for these socioeconomically backward regions and is actually important for those who do not have the wherewithal to invest in conventional irrigation systems. There are many ways productivity (both land and water) in rain-fed farming can be raised. Some of them are use of drought-resistant varieties; and the use of irrigation combined with high-yielding varieties and fertilizers and pesticides; the use of highly water-efficient and high valued dryland crops (Kumar, 2009). Sikka (2009) shows that the introduction of fishery in water harvesting pond meant for supplementary irrigation of paddy, could enhance farm returns and water productivity significantly in rain-fed paddy areas. But the poor people in these backward regions lack the knowledge and capacities to adopt the technologies needed, including the appropriate fish variety, the feed, etc. Poor knowledge about modern agricultural practices, compounded by poor information about markets and lack of marketing skills prevents them from investing in high productivity farming systems.

9.3 Constraints in irrigated agriculture

9.3.1 Physical constraints

As discussed earlier, a global review of empirical studies on water productivity shows the climate can be taken advantage of for improving water productivity for the same crop, with estimated values of water productivity for major crops viz., maize, cotton, wheat, and paddy found to be highest for locations in China, primarily due to climate, followed by on-farm water management measures (see Zwart & Bastiaanssen, 2004). Studies in the Narmada basin showed that within the same basin, great opportunities for improving water productivity of a given crop exist if we can earmark certain regions for certain crops, on the basis of the climate (Kumar, 2010). Given the fact that within South Asia, major variations in climate exist, and many crops are grown in more than one agro climate,

[a] such possibilities should be ideally significant.

But, along with water productivity, total agricultural output is also a concern for the agricultural and water sector policymakers. The regions which have favorable climate for growing a crop with less water do not seem to have sufficient land that can be allocated to the crop in question. Many water-intensive crops like paddy and wheat are today grown in regions

---

[a] For instance, sugarcane is grown in tropical and sub-tropical climates and paddy is grown in both hot and semi-arid regions and hot and humid regions.

which have large arable land but having hot and arid climate. Shifting these crops to areas with more moderate climate within the same basin or elsewhere can result in sharp decline in production, as these areas have much lower arable land, as shown by a recent analysis provided in Kumar et al., (2006) for five major river basins of India, viz., Narmada, Indus, Krishna, Sabarmati, and Cauvery. An example is growing paddy and wheat in eastern UP and Bihar instead of Punjab, Haryana, and Andhra Pradesh.

Productivity (yield per ha) is also a concern for both farmers as well as policymakers in many areas, where significant achievement in terms of water productivity gains is possible by virtue of the climate difference. For instance, the ET requirement for sugarcane is very low in the subtropical region covering UP, Bihar, and West Bengal, and the irrigation requirement further lower. Therefore, field-level WP there can be much higher than that in semiarid region of Tamil Nadu, where water requirement for sugarcane is very high—1800 mm approximately against 300–400 mm at the former region. But the yields obtained for sugarcane are one of the lowest in the former case, due to the poor solar radiation. Actual yields for many crops might be lower in those regions due to ecological reasons such as lower temperature and solar radiation which actually can reduce ET, but have negative implications for potential yield (Loomis & Connor, 1996: pp. 398).

Given the sharp variations in rainfall and climatic conditions—from hot and semiarid to Savannah to humid tropics—and renewable water availability across regions (FAO, 1986), the trade-offs associated with lower yields and poor access to arable land in regions of higher water productivity might become challenges in Sub-Saharan Africa.

9.3.2 Institutional constraints

For the same type of system, water productivity for the same crop can change at a field scale (Singh, van Dam, & Feddes, 2006: pp. 272) according to water application and fertilizer use regimes. Changing water allocation strategies at the field level can help enhance WP. For this, it is important to know the marginal productivity with respect to changing dose of irrigation water and nutrients. Farmers' water allocation decisions are governed by institutional regimes governing the use of water. Let us examine the constraints in achieving marginal productivity gains from an institutional perspective.

For a given crop, the irrigation dosage and the crop water requirement (beneficial use plus beneficial nonconsumptive use) corresponding to the maximum yield may not correspond to the maximum water productivity (Rs/m^3) (Molden, Murray-Rust, Sakthivadivel, & Makin, 2003). The WP (k/m^3) would start leveling off and decline much before the yield starts leveling off (Molden et al., 2003). Ideally, WP in terms of net return from crop per cubic meter of water (Rs/m^3) should start leveling off or decline even before physical productivity of water (kg/m^3) starts showing that trend. When water is scarce, there is a need to optimize water allocation to maximize water productivity (Rs/m^3) by changing the dosage of irrigation. But this may be at the cost of reduced yield and net return per unit of land, depending on which segment of the yield and WP response curves the current level of irrigation corresponds to.

Recent analysis with data on applied water, yield, and irrigation WP for select crops in the Narmada river basin in India showed that in many cases, trends in the productivity of irrigation water in response to irrigation did not coincide with the trends in crop yields in response to irrigation. In this case, limiting irrigation dosage might give a higher net return per unit of water. But farmers may not be interested in that unless it gives higher returns from the land. The reason is that there are no limits on volumetric water use, due to the absence of well-defined rights in surface and underground water resources. Though at the societal level, the resource might be scarce, at the individual level, the resource-rich farmers might enjoy unrestricted access to it. This is the major institutional constraint in improving water productivity (Kumar, 2009).

Hence, if the return from the land does not improve, the strategy of restricting water allocation can work only under four situations as discussed earlier (under 8.4.3).[b] The availability of extra land for expanding irrigation would be the real incentive for doing this. Under the condition of supply rationing, farmers would be using water for growing economically efficient crops anyway (Kumar, 2005; Kumar, Scott, & Singh, 2011). But the issue being addressed here is for a given crop, how far the water productivity can be enhanced to a level which the best-managed farm achieves. In all these three situations described above, the WP improvements would lead to farmers diverting the saved water for irrigating more crops to sustain or enhance their farm income. The reason is that the amount of water being handled by farmers

[b] First: the amount of water farmers can access is limited by the natural environment. Second: farmers are confronted with a high marginal cost of using water. Third: water supply is rationed (Kumar, 2009). Fourth: there is great social pressure from downstream farmers to improve the efficiency of water use, especially during droughts (Lankford, 2012).

is too small that they need to use the same quantum of water as previously used since the WP differences are just marginal (Kumar, 2009).

But, situations like those described above are not very common in South Asia, though can be encountered in large parts of SS Africa. In SS Africa, the serious technological and financial constraints in accessing groundwater (Namara, Gebregziabher, Giordano, & De Fraiture, 2013; Qureshi & Ismail, 2016) will create an incentive among farmers to use the pumped water efficiently and allocate it to crops that give high-income returns per unit of water.

In India, even in the hard rock areas with poor groundwater environments, farmers are frantically drilling boreholes to tap water from deeper strata, thereby overcoming the constraints imposed by a physical shortage. While power supply restriction is being tried by governments to limit farmers' access to groundwater, as in the case of *Jyotigram Yojna*[c] in Gujarat, in reality, this is leading to the increasing adoption of higher capacity pump sets by farmers, greater power theft through the under-reporting of connected load and more inequity in the distribution of benefits from the subsidized power supply.

There are very few locations in South Asia where canal water supply is heavily rationed in volumetric terms. In most situations, water supply is unrestricted in the command areas, though rotational water supply (*Warabandi*) system is followed in some large irrigation commands of northwestern India and the Indus system. Even in cases where rotational water supply is followed, water delivery in the field is not measured and pricing is not done on a volumetric basis. In Afghanistan, where canal irrigation accounts for nearly 75% of the total irrigation, the management of canal systems has suffered a lot due to the institutional vacuum created by wars and regional conflicts. As a result, the infrastructure for water distribution and delivery has been lying in a very dilapidated stage, with very poor irrigation performance (Qureshi, 2002).

According to FAO estimates of 1997, about 1.7 m per ha required rehabilitation and another 0.68 m per ha required improved on-farm water management conditions. Canal management had completely collapsed. Canals are completely silted, breached, and have not been delivering like in the past. About 46% of the irrigation structures are damaged and 88% of the irrigation structures are traditional which are responsible for 40% of the total water loss. This is primarily due to drought and lack of maintenance of surface

---

[c] It involved separation of feeder line for agriculture and domestic power supply.

water facilities—small dams and canal systems. The indirect impact of war on modern irrigation systems is much more serious than the traditional schemes. The intake structures of modern irrigation schemes are out of function due to the missing mechanical parts looted during the war and lack of professional staff to repair and operate these systems (Qureshi, 2002).

As regards groundwater, regulatory measures exist for controlling groundwater pumping only in a few provinces of India. However, they are not effectively enforced. The other South Asian countries haven't even made any serious attempts to control groundwater over-use. No other south Asian country has been able to enforce regulations on the volumetric withdrawal of groundwater, unlike developed countries (US, Australia, Japan, and Spain) and developing countries like China where both direct and indirect instruments for managing groundwater, viz., enforcement of tradable property rights, groundwater tax, pro-rata pricing of electricity for groundwater pumping, the introduction of volumetric quota—are effectively used. Hence, the farmers are not confronted with the opportunity cost of using water, which can create an incentive for transferring water to economically efficient uses in agriculture or use water more efficiently in physical terms or do both.

9.3.3 Market constraints

Major gains in water productivity (economic terms) are possible through crop shifts toward more water-efficient ones such as low water consuming fruits and vegetables that give high-income returns (Kumar and van Dam, 2013) at the level of individual farms, though the possibility of doing that is determined by the climate. For instance, pomegranate fruit produced in north Gujarat has applied water productivity of Rs. 39/m^3 of water under tube well irrigation under normal market conditions. But highly volatile market conditions and poor marketing infrastructure induce major constraints to improving water productivity and reducing the stress on water resources (Kumar & van Dam, 2009).

As we have discussed earlier, the demand for fruits and vegetables is increasing steadily in South Asia due to income growth (Alexandratos & Bruinsma, 2012). The food demand in South Asia is projected to increase from 2293 Kcal/person/day to 2820 Kcal/person/day, mainly because of the change in consumption pattern with more vegetables, milk, roots and tubers, and other food items, including fruits and less of cereals being consumed. Among these, fruit crops and vegetables offer great potential in terms of improving water productivity in economic terms, as they are high values market when the production is low. Another intervention is to take the produce to distant markets where the climate is not favorable for producing such crops. Provision of cold storage and instant freezing technologies are needed for this. Earmarking of large areas under traditional crops to such high-valued crops can add to the woes of the farmers. The reason being that most of these crops (many fruits and vegetables) are fast perishing, and hence need to be brought to the markets immediately after harvest.[d] Good roads are absent in many such regions, where productivity levels are very low today. In the case of crops like onion and potato, infrastructure for postharvest treatment and storage of the product would be required. In many regions in South Asia, such as eastern and northeastern India, Bangladesh, and Nepal, productivity levels are very low. These regions also lack good infrastructure including electricity and roads (Kumar, 2009).

That said, opportunities for crop diversification are unlikely to be big in Sub-Saharan Africa because a large proportion of the farmers have small holdings and there are risks associated with allocating land for commercial crops instead of food crops (Birthal, Joshi, Roy, & Thorat, 2013). The other reason is the poor (domestic) market for fruits and vegetables owing to low purchasing power. The region will have to tap the international market, especially export to Europe, for selling such high-value crops for decades till the domestic market grows as a result of rising incomes.

9.3.4 Policy constraints

Inefficient pricing of electricity in the farm sector, characterized by heavy subsidies and charging for power on the basis of connected land, is a major policy constraint to improving water productivity in agriculture in Asian countries (Kumar, 2005; Kumar, Singh, Samad, Purohit, & Didyala, 2008; Qureshi, McCornick, Sarwar, & Sharma, 2010; Zekri, 2008), though power subsidies still have not emerged as a problem threatening the sustainability of groundwater use in any of the African countries. Nearly 60% of India's irrigated area gets water supplies from wells (Kumar, 2007). Well-irrigated fields are more amenable to technologies and practices for improving crop water productivity by virtue of the greater control farmers can exercise over irrigation water application. Microirrigation is one of the most important agricultural technologies that can help improve water productivity in crops.

---

[d] Storing such produce in cold storage, etc., will not be economically viable.

Whereas control over water allocation (Kumar, Singh et al., 2008) and improving the quality and reliability of water (Kumar, Trivedi, & Singh, 2009), are the practices that can help improve water productivity. The heavy subsidies and flat-rate pricing of electricity in agriculture leave no incentive among farmers to secure higher water productivity through improved water allocation and microirrigation systems as they do not lead to improved returns from a unit area of land (Kumar & van Dam, 2009; Kumar, Singh et al., 2008).

The recent studies from the Indian states of Uttar Pradesh, Bihar, and Gujarat highlight the positive impact of introducing pro-rata pricing of electricity in agriculture on field level water productivity (Kumar, 2005 for Gujarat, India) and water productivity of the entire farming system (Kumar et al., 2011 for all the three states). Kumar, Singh, and Sivamohan (2010) showed that the price of electricity could be raised to such a level that the marginal cost of water for the electric well-owning farmer becomes equal to that of the diesel well owning farmer, provided that good quality power supply is assured (Kumar et al., 2010). But, the proposals for metering electricity in the farm sector, and metering electricity, and introducing pro-rata pricing get rejected on flimsy grounds. One of them is that farmers are rural vote banks and that raising the power tariff is highly unpopular as it would make farming less attractive. It is to an extent true that merely raising the power tariff would only lead to an increase in the cost of irrigation in areas where the power supply is of very poor quality. This is because farmers would be discouraged from choosing a cropping system that is a water-efficient, but often high risk, due to the fear of supply interruptions and crop damage. Another argument against metering and pro-rata pricing is the transaction cost of metering a large number of wells in remote rural areas.

But, one important factor that is missed out in the entire discussion on raising *power* tariffs is the improved quality of power supply possible under metered tariff. Under a flat-rate tariff, it is important to regulate the power supply to reduce the negative effects on welfare such as excessive pumping and misuse of groundwater and electricity and inequity in the distribution of subsidy benefits and greater revenue losses to the electricity board. This affects the quality of irrigation. But this is not necessary under pro-rata pricing (Kumar, 2009). Improving the quality of the power supply would change the energy-irrigation nexus (Kumar, 2005). Kumar et al. (2011) showed that under pro-rata pricing with high energy tariffs, lead to better equity in access to groundwater. Apart from securing higher water productivity, the farmers got higher returns per unit of land, used less amount of groundwater. All these are achieved through the careful selection of crops, and farming systems that use less amount of water, but give higher returns per unit of land, and use all inputs including water more efficiently.

But, as noted by Kumar (2009), these were rather excuses used by officials and other functionaries of electricity departments to cover up the revenue losses due to poor operational efficiencies, resulting from transmission losses, and distribution losses which include thefts. Also, unmetered connections attract more bribes, as detecting power theft is much more difficult under a flat-rate system. Obviously, detecting thefts like this would require field visits by the technicians, and checking the connected load. Hence, the flat-rate system is patronized by a section of the engineering staff of electricity boards. As a result, the state governments find it rather convenient to continue with such policies. Such degenerative policies act as a major constraint to improving water productivity in agriculture. But it is important to recognize the fact that resistance to metering is not from the farming lobby, but from the bureaucracy itself.

9.3.5 Technological constraints

One of the most effective ways of improving water productivity is through greater control over irrigation water delivery. In the case of well irrigation, if the electricity supply is reliable, farmers can exercise good control over water delivery. Diesel well owners in India were found to be securing high water productivity in economic terms in spite of incurring high marginal costs for irrigation water due to high diesel prices, as compared with electric pump owners who incur very low costs for using energy and water (Kumar et al., 2010). The control over irrigation water is one major factor that enables them to allocate water optimally (Kumar & Patel, 1995).

Among all South Asian countries, electricity use for groundwater irrigation is highest in India. In Pakistan, though vast areas are tube well irrigated, most of the wells are energized by diesel pumps (Qureshi et al., 2010). Though groundwater irrigation is very significant in Bangladesh with wells accounting for 75% of the total irrigated area, most of the pumps here also are run by diesel. This is in spite of the fact that diesel well irrigation is economically less attractive for farmers as compared to electric well irrigation (Zahid & Ahmed, 2006). But, farmers resort to diesel engines because of poor access to electricity in rural Bangladesh.

In Afghanistan and Nepal, the contribution of well irrigation to the total irrigated area is small (Kumar, Viswanathan, & Bassi, 2014). In India, there are very few states which offer good quality power supply to agriculture in terms of reliability and adequacy. Gujarat is one among them. Many states are facing a power crisis, and agriculture has been at the receiving end, which has to be satisfied with an irregular, erratic, untimely, and short-duration supply of electricity (Singh, 2009). Under such a supply regime, controlled and quality irrigation is not at all possible. The erratic and short-duration power supply also induces constraints in farmers adopting precision irrigation systems like drips and sprinklers which are energy-intensive in certain cases.[e]

One disincentive for well irrigators to improve crop water productivity is the lack of opportunity costs of using groundwater and electricity. Because of this, obtaining higher yield and higher net income are the primary concerns. One way of inducing this opportunity cost is through the restriction of energy use by farmers (Zekri, 2008), the other being consumption-based pricing of electricity (Kumar et al., 2011; Pfeiffer & Lin, 2013). Technologies exist for controlling energy consumption by farmers. The prepaid electronic meters, which are operated through scratch cards and that work on satellite and internet technology, are ideal for remote areas to control groundwater use online (Zekri, 2008). As Zekri (2008) notes, such technologies are particularly important when there are large numbers of agro wells, and the transaction cost of visiting wells and taking meter readings is likely to be very high. Hence, they are ideal for South Asian countries which have high well densities. But such technologies are still not accepted. Resistance to introducing such technologies due to vested interests within the state electricity departments is also notable (Kumar, 2009).

In the case of canal irrigation, devices that provide control over water delivery to the lowest part of the delivery system in the irrigation scheme are absent in most of the old gravity irrigation systems. Hydraulic structures (head regulator, cross regulators, division box), that are needed to regulate the flow and maintain food supply level in the canals across the network, are absent in most surface irrigation systems, which were designed using old design concepts. The result is that the flow and water delivery in the field is uncontrolled with fluctuations in discharge and unreliable water delivery. This is a major hindrance for farmers to exercise sufficient control over water application. Such problems of poor reliability and uncontrolled delivery occur in both water deficit and water abundant systems. While intermediate storage systems like the diggie in Rajasthan can help farmers leverage control

---

[e] This, in no way suggests that good quality power will result in farmers using efficient irrigation technology. It only means that those who are interested in opting for efficient irrigation technologies would be constrained by poor quality power supply.

over water application, in many instances they are not feasible due to problems of land availability. In the case of Bikaner in Rajasthan, Amarasinghe, Bhaduri, Singh, Ojha, and Anand (2008) showed that diggies in Rajasthan are economically viable when the landholding is larger than four acres.

In SS Africa, since groundwater irrigation is just picking up, well irrigation is mostly practiced by wealthy farmers. Wells are very few in number. A large proportion of the farmers who use energized well irrigation depend on diesel engines or petrol pumps incurring a high cost for pumping water (Namara et al., 2013). Because of this reason, overdraft of groundwater is still not an issue. As for gravity irrigation, public irrigation systems were in most cases performing below expectations. Two of the causes of poor performance in Africa included institutional inefficiencies and unreliable water supplies (Peacock, Ward, & Gambarelli, 2007). The poor performance of irrigation investments resulted from some important governance challenges relating to the way in which schemes were developed, managed, and financed (Oates, Jobbins, Mosello, & Arnold, 2015). There were multiple technical issues that the schemes face as a result of poor engineering and maintenance (Palmer-Jones, 1987). Many of the agencies administering large schemes were established primarily to deal with construction. They were illequipped for ongoing management, and incentives to provide reliable services to farmers were often absent (Oates et al., 2015). Experience suggests that agricultural productivity improvement through the modernization of irrigation systems does exist in SS Africa, though past outcomes were varied (Oates et al., 2015).

9.3.6 Socioeconomic constraints

The larger social objective of improving national food security would continue to act as a major constraint to improving water productivity in economic terms in South Asia and SS Africa, both being highly food-insecure regions. A nation, which faces water shortages, may still decide to produce food within its own territory instead of resorting to imports due to political and economic reasons, rather than importing food from regions that can produce it at lower costs or with higher resource use efficiency. For instance, Indian Punjab produces surplus wheat and rice and supplies them to many other parts of India, which are food-deficit, including eastern India (Amarasinghe et al., 2004; Kumar, Gulati, & Cummings, 2007). Twenty percent of the country's wheat production and 10% of its rice production came from Punjab; it contributed 57% and 34%, respectively, to the central pool of grains for public distribution (Kumar et al., 2007).

In such situations, economic efficiency of production may not be of great relevance, as the country may strategically decide to subsidize the farmers to procure systems that provide control over irrigation water delivery thereby improving the physical productivity of water. Instead, the opportunity cost of importing food may be taken into consideration and treated as an additional benefit while estimating the incremental value of crop output for estimating the "economic productivity" function. These considerations seem to be extremely important for countries of South Asia and Sub-Saharan Africa. For instance, the analysis by Shah and Kumar (2008) shows that the social benefits due to incremental food production (42 million tons) through investment in large reservoir projects in India since Independence, in terms of lowering food prices, was to the tune of Rs. 4290 crore annually. If the food security impacts of producing a certain crop are considered in analyzing crop water productivity, then at least the arguments of low water productivity in gravity irrigated paddy-wheat systems would become invalid.

But, as noted by Kumar and van Dam (2009), food security cannot be an excuse for low productivity. Many canal irrigation systems, in fact, suffer from poor quality and reliability of water supplies, which actually affects water productivity adversely. Therefore, the attempt should be to improve the productivity of the water by improving the quality and reliability of water supplies in canal irrigation systems that mimic well irrigation. This assumes greater significance when one considers the fact that most of the irrigated cereals such as paddy, wheat, bajra, and jowar have much lower water productivity in simple economic terms as compared to oilseeds, vegetables, and fruits (Kumar & van Dam, 2009).

Labor absorption capacity of irrigated agriculture is another socioeconomic consideration in deciding agricultural productivity enhancement measures. Paddy is labor intensive, and a large chunk of the migrant laborers from Bihar work in the paddy fields of Indian Punjab. As per some estimates, 2.614 million ha of irrigated paddy in Punjab (as per 2005 estimates) create 159 million labor days[f] during the peak Kharif season. The total percentage of farm labor contributed by migrant laborers during peak season was reported to be 35% as per the Economic Survey of Punjab 1999–2000 (GOP, 2001). The total number of labor days contributed by migrant laborers to paddy fields in Punjab was estimated to be 55.75 million

[f] This is based on the primary data which show that a hectare of paddy creates Rs. 5000 worth of farm labor in Punjab. This is exclusive of the machinery employed in plowing and harvesting. With a labor charge at the rate of Rs. 80 per day, the number of labor days per ha of irrigated paddy is estimated to be 61 (Source: primary data from Punjab).

(Singh & Kumar, 2009). Hence, replacing paddy by cash crops would mean a reduction in farm employment opportunities (Kumar & van Dam, 2009). Hence, this can be a constraint in many backward regions of South Asia where labor is in surplus, including large parts of Bangladesh and Eastern India. However, the reduction in manual labor requirements in farming would be a boon in many agriculturally prosperous regions of India and Sri Lanka.

Conflicts and war had a remarkable impact on agricultural productivity in Afghanistan, through damage to the country's irrigation infrastructure. But this damage was less a consequence of destruction caused by war but much more a consequence of the migration of farmers to other countries leaving behind the irrigation schemes unattended. Farmers have abandoned about 40% of the land due to a lack of maintenance and 10% of the land is completely destroyed due to war. About one-half of all irrigation schemes were in need of rehabilitation. Surface water irrigation systems are performing less than half of the 1980 levels. As a result, irrigated area has drastically reduced. As per the estimates of the irrigation department, only 1.4 m per ha of land was irrigated in 2002, which was half that of 1980 levels. Afghan farmers use farming techniques that are centuries old (Qureshi, 2002).

Little knowledge on new irrigation technologies and cultivation practices is available to the farmers, and institutional credit facilities are almost absent. As a result, irrigation system efficiencies are rated to 25–30% and productivity levels are low even by regional standards. About 20% of all irrigation systems required improvements in on-farm water management to avoid underirrigation (low crop yields) or over-irrigation (waterlogging and salinization).

9.3.7 Environmental constraints

Environmental system considerations can induce new constraints in the use of approaches for improving water productivity in irrigated production. Well irrigation exists in large canal commands, in lieu of the poor reliability of canal water supplies. Examples are Sindh, Baluchistan, and Punjab provinces covered by IBIS in Pakistan and Indian Punjab. The area irrigated by tube wells has surpassed the area irrigated by canals over the years in India (source: based on Kumar, 2007); whereas in the IBIS, the contribution of well irrigation progressively increased from 11.3 BCM in 1965–1966 to 52 BCM in 1996–1997, while water diversion through canals changed only marginally (Source: based on fig. 1 in Khan, McCornick, & Khan, 2008). Interestingly, in many situations, well irrigation is sustained by return flows from canal seepage and irrigated fields.[g] In Japan, irrigated paddy is promoted using the concept of PES (payment for ecosystem services) to improve groundwater recharge in areas where the large-scale reduction in irrigated paddy cultivation led to reduced recharge and depletion of groundwater (OECD, 2017). Such return flows are significant in irrigated paddy (Kim, Jang, Im, & Park, 2009; Singh, 2005), which occupy a significant chunk of the irrigated area in most canal commands in South Asia, particularly in the two examples cited above. Therefore, an improvement in the water productivity of crops grown under gravity irrigation through technical efficiency of irrigation (such as SRI, alternate wetting, and drying for paddy and microirrigation systems for other crops) would be at the cost of sustainability of well irrigation.

Such concerns of maintaining return flows from irrigated fields to shallow aquifers will be relevant for SS Africa too, given the fact that groundwater potential is extremely limited in most parts due to crystalline formations, and that recharge from irrigated fields in the form of return flows can incentivize farmers to invest in drilling wells for irrigation.

Whereas in certain other cases, well irrigation prevents conditions of water logging which can occur due to excessive seepage and return flows from canals as seen in Ukai–Kakrapar and Mahi command in South and Central Gujarat, respectively in India, and IBIS in Pakistan. In the first case, while analyzing the productivity impacts of canal irrigation, the environmental benefit of providing the critical recharge that protects the groundwater environment can be considered in the numerator of water productivity. In the second case, while analyzing the water productivity impact of well irrigation, positive environmental effects of preventing water logging conditions should be considered in assessing the net returns, while there may not be any opportunity costs associated with its use for irrigation. Here again, a reduced depletion of groundwater, which might be the outcome of efficiency improvements in irrigation of the crops watered by well water or a mere reduction in groundwater pumping due to high costs, etc., can lead to more environmental problems. These concerns are real. For instance, an increase in electricity tariff for agricultural use and diesel cost in Pakistan resulted in the large reduction in groundwater withdrawal in the IBIS during 1997–1998, in the order of 13.6 BCM (Khan et al., 2008).

---

[g] This is dependent on soil conditions and geo-hydrological settings. There are numerous examples of this phenomenon. To cite a few are IBIS, known as one of the largest water recycling machines in the world; irrigation system in Indian Punjab and Mula irrigation command in Maharashtra, western India.

9.4 Conclusions

In this chapter, we have discussed the constraints in improving water productivity in rainfed and irrigated agriculture separately. In the case of rainfed agriculture, the constraints are social, socioeconomic, and financial in nature. As regards the first one, the poor people in the backward regions where water and land productivity are poor, lack the knowledge about modern agricultural practices and capacities to adopt the technologies. This is applicable to both South Asia and SS Africa. This, compounded by poor information about markets and a lack of marketing skills, prevents them from investing in high productivity farming systems. The regions of south Asia which are water-rich have very high concentrations of native tribal people who are used to growing subsistence crops that have low economic returns and water productivity. This is the same with the humid regions of SS Africa. Hence, most of the preconditions for achieving water productivity gain through supplementary irrigation are not likely to be met. Investments for water harvesting and groundwater recharge schemes that can help improve water productivity in rain-fed farming systems are very high when compared to the amount of water they can provide, even if they produce high economic and social returns. The poor farmers of SS Africa and the people in the tribal-dominated hill regions of India are least likely to mobilize these resources. Hence, there are financial constraints too. Large-scale government financing of water harvesting and groundwater recharge systems would therefore be required.

As regards irrigated agriculture, the constraints are many more. They are physical, technology-related, socioeconomic, policy-related, institutional, market-related, and environmental. As regards physical constraints, in many regions of South Asia, where crops can be grown with less amount of water (low ET by virtue of the climatic factors), and the arable land availability is very poor. Further, the yield levels for major crops are also much lower compared to regions that are arider. The lack of ability to introduce technologies that can restrict the access of millions of farmers to water and electricity is a technology constraint for improving water productivity.

Inefficient pricing of water and electricity supplied for irrigation poses a major policy constraint in most south Asian countries, though not a major concern for SS Africa where gravity irrigation is extremely limited and private well irrigators largely use their own energy sources. In the absence of an effective institutional regime for restricting farmers' access to groundwater for irrigation in most situations (in south Asia), the interest of the farmers remains in maximizing the returns per unit of land, and not maximizing the returns per unit of water, the latter being the societal interest when the resource is scarce. The irrigation corresponding to a high return per unit of land (which is the private interest) is usually much less than that which corresponds to high returns per unit of water, which is the societal interest. While major gains in water productivity are possible through a shift to high-value crops, they pose significant market risks due to the seasonal nature of many fruits and vegetables, oversupply of produce during the season, poor adoption of postharvest technologies, and lack of transportation and market infrastructure. Finally, addressing environmental concerns of improving groundwater sustainability in certain cases, and preventing waterlogging in certain other cases also limit our options to improve water productivity in irrigated agriculture, the latter being relevant for the South Asian region.

References

Alexandratos, N., & Bruinsma, J. (2012). World agriculture towards 2030/2050: The 2012 revision. ESA Working Paper # 12–03, Agricultural Economics Development Division, Food and Agriculture Organization of the United Nations, June 2012.

Amarasinghe, U., Bhaduri, A., Singh, O. P., Ojha, A., & Anand, B. K. (2008). Cost and benefits of intermediate water storage structures: Case study of Diggies in Rajasthan. In M. D. Kumar (Ed.), *Vol. 1. Managing water in the face of growing scarcity, inequity and declining returns: Exploring fresh approaches. Proceedings of the 7th Annual Partners Meet, IWMI TATA Water Policy Research Program, ICRISAT, Patancheru, Hyderabad, India* (pp. 51–66). 2–4 April 2008.

Amarasinghe, U., Sharma, B. R., Aloysius, N., Scott, C., Smakhtin, V., & De Fraiture, C. (2004). *Spatial variation in water supply and demand across river basins of India, research report 83*. International Water Management Institute.

Birthal, P. S., Joshi, P. K., Roy, D., & Thorat, A. (2013). Diversification in Indian agriculture toward high-value crops: The role of small farmers. *Canadian Journal of Agricultural Economics, 61*, 61–91.

Food and Agriculture Organization of the United Nations. (1986). *Irrigation in Africa south of the Sahara*. FAO Investment Centre Technical Paper 5. Rome, Italy: Food and Agriculture Organization.

Food and Agriculture Organization of the United Nations. (2011). *The state of world's land and water resources for food and agriculture: Managing systems at risk*. Rome/London: FAO/Routledge.

Food and Agriculture Organization [FAO]. (2015). *Food and Agriculture Organization of the United Nations*. Available at: http://www.fao.org/3/a-bc600e.pdf.

Government of Punjab. (2001). *Economic Survey of Punjab 1999–2000*. Economic Adviser to Punjab Government, Government of Punjab, Chandigarh.

Khan, A. H., McCornick, P., & Khan, A. R. (2008). Evolution of managing water for agriculture in the Indus River Basin. In *CGIAR challenge program on water and food* (p. 120).

Kim, H. K., Jang, T. I., Im, S. J., & Park, S. W. (2009). Estimation of irrigation return flow from paddy fields considering the soil moisture. *Agricultural Water Management, 96*(5), 875–882.

Kumar, A. G., Gulati, A., & Cumming, R. (2007). *Foodgrains policy and management in India. Responding to today's challenges and opportunities.* Washington, DC: IFPRI.

Kumar, M. D. (2005). Impact of electricity prices and volumetric water allocation on energy and groundwater demand management: Analysis from Western India. *Energy Policy,* *33*(1), 39–51.

Kumar, M. D. (2007). *Groundwater management in India: Physical, institutional and policy alternatives.* New Delhi, India: Sage Publications.

Kumar, M. D. (2009). Opportunities and constraints in improving water productivity in India. In M. D. Kumar, & U. Amarasinghe (Eds.), *Strategic Analyses of the National River Linking Project (NRLP) of India series 5. Water productivity improvements in Indian agriculture: Potentials, constraints and prospects* (p. 163).

Kumar, M. D. (2010). *Managing water in river basins: Hydrology, economics, and institutions.* New Delhi: Oxford University Press.

Kumar, M. D., Ghosh, S., Patel, A., Singh, O. P., & Ravindranath, R. (2006). Rainwater harvesting in India: Some critical issues for basin planning and research. *Land Use and Water Resources Research,* *6*(1732–2016–140267).

Kumar, M. D., & Patel, P. J. (1995). Depleting buffer and farmers response: Study of villages in Kheralu, Mehsana, Gujarat. In M. Moench (Ed.), *Electricity prices: A tool for groundwater Management in India? Monograph.* VIKSAT-Natural Heritage Institute: Ahmedabad.

Kumar, M. D., Patel, A., Ravindranath, R., & Singh, O. P. (2008). Chasing a mirage: Water harvesting and artificial recharge in naturally water-scarce regions. *Economic and Political Weekly,* 61–71.

Kumar, M. D., Reddy, V. R., Narayanamoorthy, A., Bassi, N., & James, A. J. (2018). Rainfed areas: Poor definition and flawed solutions. *International Journal of Water Resources Development,* *34*(2), 278–291.

Kumar, M. D., Scott, C. A., & Singh, O. P. (2011). Inducing the shift from flat-rate or free agricultural power to metered supply: Implications for groundwater depletion and power sector viability in India. *Journal of Hydrology,* *409*(1–2), 382–394.

Kumar, M. D., Singh, O. P., Samad, M., Purohit, C., & Didyala, M. S. (2008). Water productivity of irrigated agriculture in India: Potential areas for improvement. In *Proceedings of the 7th Annual Partners' Meet of IWMI-Tata Water Policy Research Program.* International Water Management Institute, South Asia Regional Office, ICRISAT Campus, Hyderabad, 2–4 April 2008.

Kumar, M. D., Singh, O. P., & Sivamohan, M. V. K. (2010). Have diesel price hikes actually led to farmer distress in India? *Water International,* *35*(3), 270–284.

Kumar, M. D., Trivedi, K., & Singh, O. P. (2009). Analyzing the impact of quality and reliability of irrigation water on crop water productivity using an irrigation quality index. In M. D. Kumar, & U. Amarasinghe (Eds.), *Strategic Analyses of the National River Linking Project (NRLP) of India Series 5. Water productivity improvements in Indian agriculture: Potentials, constraints and prospects* (p. 163).

Kumar, M. D., & van Dam, J. C. (2009). Improving water productivity in agriculture in India: Beyond 'more crop per drop. In M. D. Kumar, & U. Amarasinghe (Eds.), *Strategic analyses of the National River Linking Project (NRLP) of India series 5. Water productivity improvements in Indian agriculture: Potentials, constraints and prospects* (p. 163).

Kumar, M. D., & van Dam, J. C. (2013). Drivers of change in agricultural water productivity and its improvement at basin scale in developing economies. *Water International,* *38*(3), 312–325.

Kumar, M. D., Viswanathan, P. K., & Bassi, N. (2014). Water scarcity and pollution in South and Southeast Asia: Problems and challenges (Chapter 12). In P. G. Harris & G. Lang (Eds.), *Routledge Handbook of Environment and Society in Asia* (pp. 197–215). London: Routledge, Taylor & Francis Group.

Lankford, B. (2012). Fictions, fractions, factorials and fractures; on the framing of irrigation efficiency. *Agricultural Water Management, 108*(2012), 27–38.

Loomis, R. S., & Connor, D. J. (1996). *Productivity and management in agricultural systems.* Cambridge, New York and Australia: Cambridge University Press.

Macauley, H. (2015). Cereal crops: Rice, maize, millet, sorghum, wheat. In *Background paper prepared for the Int. conference "feeding Africa", Abdou Diouf international conference centre, Dahar, Senegal.* 21–23 October 2015.

Molden, D., Murray-Rust, H., Sakthivadivel, R., & Makin, I. (2003). A water-productivity framework for understanding and action. In J. W. Kijne, R. Barker, & D. J. Molden (Eds.), *Vol. 1. Water productivity in agriculture: Limits and opportunities for improvement.* Cabi.

Namara, R. E., Gebregziabher, G., Giordano, M., & De Fraiture, C. (2013). Small pumps and poor farmers in sub-Saharan Africa: An assessment of current extent of use and poverty outreach. *Water International, 38*(6), 827–839.

Oates, N., Jobbins, G., Mosello, B., & Arnold, J. (2015). *Pathways for irrigation development in Africa: Insights from Ethiopia, Morocco and Mozambique, future agricultures.* Working Paper 119.

OECD. (2017). Payments for groundwater recharge to ensure groundwater supply in Kumamoto, Japan. In *Groundwater allocation: Managing growing pressures on quantity and quality.* Paris: OECD Publishing. https://doi.org/10.1787/9789264281554-9-en.

Palmer-Jones, R. (1987). Irrigation and the politics of agricultural development in Nigeria. In M. J. Watts (Ed.), *State, oil and agriculture in Nigeria.* Berkeley CA, USA: University of California Press.

Peacock, T., Ward, C., & Gambarelli, G. (2007). *Investment in agricultural water for poverty reduction and economic growth in Sub-Saharan Africa.* Synthesis Report Washington DC, USA: The World Bank.

Pfeiffer, L., & Lin, C. (2013). *The effects of energy prices on groundwater extraction in agriculture in the High Plains Aquifer (No. 328-2016-12730).*

Qureshi, A. S. (2002). *Water resources management in Afghanistan: The issues and options.* Research Report # 49 Colombo, Sri Lanka: International Water Management Institute.

Qureshi, A. S., & Ismail, S. (2016). Improving agricultural productivity by promoting low-cost irrigation technologies in Sub-Saharan Africa. *Global Advanced Research Journal of Agricultural Science, 5*(7), 283–292.

Qureshi, A. S., McCornick, P. G., Sarwar, A., & Sharma, B. R. (2010). Challenges and prospects of sustainable groundwater management in the Indus Basin, Pakistan. *Water Resources Management, 24*(8), 1551–1569.

Rao, C. S., Gopinath, K. A., Rao, C. R., Raju, B. M. K., Rejani, R., Venkatesh, G., & Kumari, V. V. (2016). Dryland agriculture in South Asia: Experiences, challenges and opportunities. In M. Farooq & K. H. Siddique (Eds.), *Innovations in dryland agriculture.* Springer.

Rockström, J., Barron, J., & Fox, P. (2002). Rainwater management for improving productivity among small holder farmers in drought prone environments. *Physics and Chemistry of the Earth, 27*(2002), 949–959.

Rockström, J., Karlberg, L., Wani, S. P., Barron, J., Hatibu, N., Oweis, T., … Qiang, Z. (2010). Managing water in rainfed agriculture—The need for a paradigm shift. *Agricultural Water Management, 97*(4), 543–550.

Scott, C. A., Silva-Ochoa, P., Florencio-Cruz, V., & Wester, P. (2001). Competition for water in the Lerma-Chapala basin. In *The Lerma-Chapala watershed* (pp. 291–323). Boston, MA: Springer.

Shah, Z., & Kumar, M. D. (2008). In the midst of the large dam controversy: Objectives, criteria for assessing large water storages in the developing world. *Water Resources Management, 22*(12), 1799–1824.

Sikka, A. K. (2009). Water productivity of different agricultural systems. In M. D. Kumar & U. Amarasinghe (Eds.), *Water productivity improvements in Indian agriculture: Potentials, constraints and prospects* (p. 163). Strategic Analyses of the National River Linking Project (NRLP) of India Series 4.

Singh, R. (2005). *Water productivity analysis from field to regional scale.* Doctoral thesis Wageningen, The Netherlands: Wageningen University. with references - with summaries in English and Dutch.

Singh, A. (2009). *A policy for improving efficiency of agriculture pump sets in India: Drivers, barriers and indicators* (September 1, 2009). Climate Strategies.

Singh, O. P., & Kumar, M. D. (2009). Impact of dairy farming on agricultural water productivity and irrigation water use. In M. D. Kumar & U. Amarasinghe (Eds.), *Water productivity improvements in Indian agriculture: Potentials, constraints and prospects* (pp. 85–98). Strategic Analyses of the National River Linking Project (NRLP) of India Series 4.

Singh, R., van Dam, J. C., & Feddes, R. A. (2006). Water productivity analysis of irrigated crops in Sirsa district, India. *Agricultural Water Management, 82*(3), 253–278.

Yaduvanshi, N. P. S., Setter, T. L., Sharma, S. K., Singh, K. N., & Kulshreshtha, N. (2012). Influence of waterlogging on yield of wheat (*Triticum aestivum*), redox potentials, and concentrations of microelements in different soils in India and Australia. *Soil Research, 50*(6), 489–499.

Zahid, A., & Ahmed, S. R. U. (2006). *Groundwater resources development in Bangladesh: Contribution to irrigation for food security and constraints to sustainability.* Dhaka, Bangladesh: Ground Water Hydrology Division, Bangladesh Water Development Board.

Zekri, S. (2008). Using economic incentives and regulations to reduce seawater intrusion in the Batinah coastal area of Oman. *Agricultural Water Management, 95*(3), 243–252.

Zwart, S. J., & Bastiaanssen, W. G. M. (2004). Review of measured water productivity values for irrigated wheat, rice, cotton and maize. *Agricultural Water Management, 69*(2004), 115.

CHAPTER 10

Framework for identifying the interventions required for enhancing water productivity at various scales

10.1 Introduction

There are important reasons to improve agricultural water productivity. They are: meeting the rising demand for food from a growing, wealthier, and increasingly urbanized population, in light of water scarcity; responding to pressures to reallocate water from agriculture to cities, and ensuring that water is available for environmental uses; contribute to poverty reduction and economic growth. For the rural poor, more productive use of water can mean better nutrition for families, more income, productive employment, and greater equity (Molden et al., 2007), if the availability of water in relation to cultivable land is limited. Targeting high water productivity can reduce investment costs by reducing the amount of water that has to be withdrawn to meet the agricultural water demands, but the same can make economic sense only if the investment cost for saving water through measures that improve water productivity is less than the investment required.

Molden et al. (2007) had identified four high-priority areas for water productivity gains. They are (1) areas where poverty is high and water productivity is low, where improvements could particularly benefit the poor, as in much of Sub-Saharan Africa and parts of South Asia and Latin America; (2) areas of physical water scarcity where there is intense competition for water, such as the Aral Sea Basin and the Yellow River, especially where gains in economic water productivity are possible; (3) areas with little water resources development where high returns from a little water can make a big difference; and (4) areas of water-driven ecosystem degradation, such as falling groundwater tables, river desiccation, and intense competition for water.

This framework responds to the broader concerns related to agricultural growth in areas facing resource depletion and competition for water. But it fails to characterize regions that are water-abundant, yet experience an agricultural stagnation, low rural economic growth, and high incidence of poverty, from an economic perspective. In the case of water-abundant regions, the extent to which agricultural water productivity improvement can help raise farm outputs, fuel rural growth and reduce poverty depends largely on the availability of cultivable land that is able to utilize that water. Water productivity improvements can translate into greater agricultural outputs only when water availability in relation to cultivable land is low and there is a sufficient amount of cultivable land.

Many regions in the world are water-abundant (Kumar & Singh, 2005) and land-scarce (Fischer, Hizsnyik, Prieler, & Wiberg, 2011), and scarce in both land and water (Kumar, Bassi, & Singh, 2020). Many countries in South and South East Asia and Central Africa belong to the former, and some countries in SS Africa belong to the latter (Kumar et al., 2020). Investing in water productivity in such regions will not make much sense. Hence, in order to understand the need for and the opportunities and constraints of improving water productivity and to know where such measures would yield benefits (in terms of an increase in agricultural outputs, rural growth, and poverty reduction), we need a better framework that characterizes the river basins. Such a framework should look at the following attributes: renewable water available in the basin, how much of it is appropriated for meeting the existing demands, including agriculture, what proportion of the current demands are met from the current supply potential and whether there are unmet demands in the basin which can be met from further exploitation of the resources.

In the next section, we will illustrate how some of the measures implemented in the past to improve the water use efficiency and water productivity in agriculture, which were based on a limited understanding of the characteristics of the water resource system dynamic, were ineffective. In the subsequent section, we will present an analytical framework that helps characterize the river basins on the basis of the attributes mentioned earlier, and identify the water productivity improvement measures that are possible at various scales, i.e., plant-level, plot-level, irrigation system level, and basin level.

10.2 Some flawed solutions of the past

In this section, we will discuss some of the most touted, yet flawed schemes implemented in the past in some South Asian countries to improve water productivity or to save water in agriculture, yet being replicated in development projects due to their continuing popularity in policy circles in the absence of any objective evaluation of their impacts. Their popularity is evident from the fact that they are used as a straight-jacket solution in every irrigation modernization project, irrespective of the characteristics of the water resources, and climatic and environmental conditions that exist in the project locations.

10.2.1 Canal lining

The rationale for canal lining in irrigation projects is improving the conveyance efficiency in irrigation (Ali, 2011; Jadhav, Thokal, Mane, Bhange, & Kale, 2014). The underlying premise is that a significant portion of the water being supplied from the source is lost in seepage. In the water-related project and policy circles, mentioning about poor efficiency in public irrigation systems is in vogue. Often the figure quoted for India, which has nearly 22 m per ha of net canal irrigated area, in planning documents (like the erstwhile Planning Commission of India) is "35 to 40 per cent" (see, CWC, 2008). No one really knows the source of these figures. Though there were studies conducted in some irrigation project commands in South India, which show a very high conveyance loss in the main, branch, and distributory canals, those studies simply assumed that the water that is lost in the conveyance is not retrieved. However, this is not the case. Groundwater irrigation is well developed in canal command areas, even in high rainfall regions (like in the Damodar valley project command in West Bengal). A significant portion of the water that seeps through the unlined canal reaches the groundwater table (Kumar & van Dam, 2013; Perry, 2007), depending on the depth to the water table, climatic conditions, and soils,[a] and can be recycled back using wells. Therefore, such assumptions are invalid (Kumar & van Dam, 2013).

A proper determination of real water losses through unlined canals needs to be undertaken before taking a decision on canal lining that requires a significant investment (Sheng, King, Aristizabal, & Davis, 2003). Canal lining as a water management intervention can be justified only if the opportunity

[a] It will increase with decreasing depth to water table, increase in hydraulic conductivity of the soil and increase in aridity (source: based on Kumar & van Dam, 2013; Watt, 2008).

cost of not lining the canal exceeds the cost of the lining. Let us assume that X amount of water is lost in conveyance, and Y is recovered through wells. In that case, the amount of water that cannot be retrieved is (X–Y) and this is the real water saving in quantitative terms. Here, the opportunity cost of not canal lining (or the benefit of lining) is the sum of the incremental income from crop production that can be generated from the use of water that can be "saved" through lining (X–Y) plus the cost of obtaining the water that is retrievable (Y). This means, if all that water is recoverable or recovered, then the benefit will be limited merely to the cost of pumping that X amount of water back, which can be avoided through the lining. This cost can be quite low if the water table in the area is shallow, which is generally the case in canal commands. Here again, the cost can include only the variable cost of pumping (i.e., the energy cost), as farmers might have already invested in wells in and around the command area for pumping water.

However, if no fraction of the water that is lost in seepage is retrievable (due to it going into a saline aquifer) or retrieved (due to farmers not having wells or since well irrigation is legally not permitted like in Afghanistan and some of the countries in the Middle East), then the benefit can be quite significant, and will be equal to the incremental income from crop production using that water (=Y X Ø). Here, Ø is the economic value of the water in $/m^3, which is the surplus-value product generated from the use of water. Here again, the assumption is that the seepage from canals is completely stopped through the lining. Such an assumption can be tenable only if the lining is made of cement concrete which is very expensive. In reality, the conveyance efficiency of even the lined canals may decline over time due to material deterioration and poor maintenance (Howell, 2003).

However, what is described above is the simplest and perhaps the most ideal situation. All these benefits of lining will be nonexistent in a situation wherein in certain seasons, there is a large amount of water in the scheme. In such situations, the opportunity cost of not lining is almost zero for those seasons. This is generally the case of many diversion-based irrigation schemes in monsoon Asia. The flow in the rivers even in the semiarid and arid areas is highly variable, with 90%–95% of the total flow occurring during the monsoon in the rainfed rivers, and during summer in the snow-fed rivers.

The benefits of lining in these diversion-based schemes will be affected only for a short period of time when the flow is very low and farmers in the command area still cultivate crops. This is a situation in hundreds of thousands of traditional irrigation systems in Afghanistan, and many of the large diversion-based irrigation systems in the Ganga–Brahmaputra–Meghna basin (http://www.fao.org/nr/water/aquastat/countries_regions/profile_segments/gbm-IrrDr_eng.stm). Since the flows during the lean season are low, the amount of water that can be saved will also below. In Afghanistan, the demand for water during much of the low flow season (September to February) is very low, and it is only for 5 to 6 months (i.e., during June to November in some cases and July to December in some others) when the demand remains high and the supplies available from the canals that off-take from the natural streams are low (World Bank, 2010). In large parts of the Ganges basin, plenty of groundwater and surface water resources are available (CGWB, 2014; CWC, 2017). The water availability for irrigation during summer is not dependent on releases from canals, and farmers can get enough water at low costs from wells. In such situations, discharge measurements at the head, middle and tail reach of the canals will have to be undertaken during different seasons to quantify the amount of water lost in seepage and that which can be put to beneficial use.

The worst-case scenario, which can work against any decision to do canal lining, however, is when the high flow season corresponds to a season of low irrigation demand, and the seepage during the season augments groundwater recharge which is used by farmers during the lean season. The point here is that the water saved through lining during the high flow season will not have much value in a diversion-based scheme or a reservoir-based scheme that doesn't have an adequate storage capacity to keep the water for the lean season. As a matter of fact, groundwater storage (banking) can be a viable alternative to increasing the reservoir storage capacity in regions that experience highly variable annual flows, if recharging through canals is well managed.

Overall, the economics of lining will be sound if the volume of water lost through seepage (which is not recoverable) per unit area and the value of that water ($\emptyset$) high, and the cost per unit area of lining is low. Considering all these factors, it appears that the benefits of canal lining will be affected only in a few regions/localities where the surface water is scarce and groundwater irrigation is either not feasible or the seepage is not retrievable (like in deep groundwater areas and areas underlain by saline aquifers). The foregoing analysis suggests a careful evaluation of the benefits and costs before we indulge in canal lining projects.

However, in none of the canal rehabilitation or irrigation modernization projects, such analyses were ever undertaken. The view of the irrigation project officials of large irrigation schemes is myopic that they do not consider groundwater irrigation in the command areas as beneficial or complementary to gravity irrigation. In a recent World-Bank supported the project in the Damodar valley command in West Bengal, canal lining was proposed for spreading the limited water available during the lean (summer) season to more areas, with the assumption that groundwater that otherwise would be used to supplement canal water by the tail end farmers could be saved to prevent its depletion. The point is that in the Damodar valley command, groundwater is available in plenty and all the seepage directly contributes to the recharge of the shallow aquifer that is being tapped by farmers for irrigation. Any intervention to line the canal (to save the water during the lean season) would only reduce recharge in the command (while pumping is reduced), thereby resulting in no change in the groundwater balance. Similar inferences were drawn in the Uda Walawe irrigation scheme in southern Sri Lanka where it was estimated that after the concrete lining, the annual groundwater recharge in the command areas would reduce by about 50% (Meijer, Boelee, Augustijn, & van der Molden, 2006). Groundwater depletion and equity in water distribution are merely being used as a ploy to justify canal lining.

10.2.2 Using efficient irrigation methods in paddy

In several countries of South Asia, paddy is being targeted as a water-intensive crop in terms of the amount of water used to produce a kg of rice, responsible for unsustainable water use, especially groundwater use. Some of the recent development projects that are aimed at developing and promoting climate-resilient agricultural practices in the South Asian region, with bilateral aid, have strong components to reduce the area under irrigated paddy cultivation. Groundwater depletion in Indian and Pakistan Punjab has long been attributed to irrigated paddy in the two regions. The basis for such a view is in the amount of irrigation water applied to paddy, which is often quite large in the low rainfall regions like Indian and Pakistan Punjab.

It is also being argued that the conventional method of growing paddy under partially submerged conditions results in huge wastage of water, and water use can be significantly reduced if the crop is grown scientifically. A science magazine reported scientists from the Indian Agricultural Research Institute as saying that not more than 600 liter of water is needed to produce 1 kg of paddy, if proper water management techniques are followed in paddy cultivation, while according to the magazine, the current water use is 25 times this quantity (i.e., 15,000 liter) (Down to Earth, 2015). Such mind-boggling numbers of water consumption in traditional paddy growing, and extent of saving through new techniques such as system of rice intensification (SRI), zero tillage, alternate wetting and drying (AWD), etc. are regularly being reported in Indian media without having the support of any scientific data (see, Down to Earth, 2015).

Proposals are also made to promote water-efficient ways of paddy growing in Indian Punjab, which involve incentives to farmers for limiting energy consumption for running pumps that are used for irrigating the crop, thereby limiting water application (Chaba, 2019).[b] The basic premise for such proposals is that farmers apply irrigation water excessively because both water and electricity are made available for free, which is consumed nonbeneficially.

There is absolutely no doubt that the irrigation water requirement for the conventional method of growing paddy is quite high. However, there is a considerable difference between water applied and consumptive water use in the case of paddy (Ahmad, Masih, & Turral, 2004; Singh, 2005). Not all that water applied in the paddy field is consumed by the plant. A significant part of this water is available for reuse and recycled back through wells. Such facts are quite well understood by scientists and technocrats (see Hira & Khera, 2000).

Studies from different parts of the world, including Pakistan Punjab, Haryana in India, and South Korea show that percolation of water from irrigated water to shallow groundwater is a major component of the total water applied to the paddy field, and it will be wrong to reconsider this water as depleted or lost. For instance, the analysis of soil water balance in two rice-wheat fields in Sirsa district of Haryana (which had a similar climate) using SWAP model showed that the total water applied (1239 and 1427 mm, respectively) was in excess of the estimated ET (949 mm and 858 m) in the order of 290 and 561 mm. This excess water was available as recharge to groundwater and soil moisture storage change. Interestingly, the actual ET value was higher for the field which had a lower dosage of irrigation (Singh, 2005). Field measurements undertaken in South Korea showed a percolation of around 1220 mm and another 557 mm as surface runoff,

---

[b] This model involves offering cash incentives to well irrigator farmers in Punjab who use less than designated quota of electricity each season. The quota for each farmer is decided on the basis of the connected load he/she had and the season. For one HP of connected load, a farmer is entitled to 200 units of power per month during the kharif season and 50 units during the winter season (Prasad, 2018). The assumption here is that paddy is four times more water guzzling than wheat in Punjab. If a farmer manages to limit the energy consumption to a level which is 1000 units less than the quota, he/she would be entitled to get a cash incentive of Rs. 4000 (Rs. 4 per unit of electricity saved). Consumption above this quota won't invite any penalty though (Chaba, 2019).

out of the total of 2575 mm of water applied in the paddy fields (irrigation and rainfall). The ET was only 798 mm (Kim, Jang, Im, & Park, 2009). While paddy is grown mostly during the monsoon season in many parts of Asia, paddy fields act as a spreading basin for rainwater to infiltrate and help improve the groundwater balance.

One might argue that a reduction in soil moisture depletion is one avenue for saving groundwater, but this fraction of water use is unlikely to be significant (as farmers in Punjab go for wheat cultivation soon after harvest of paddy, some of the residual moisture might get beneficially used). Hence, to sum up, even if the farmers reduce the amount of water applied to irrigated paddy fields, there won't be any groundwater savings.

A meta-analysis of studies looking at the impact of the AWD method of paddy irrigation, which involved a total of 58 studies involving 528 side-by-side comparisons, showed a negative impact of AWD on yield (5.4%) and a reduction in water use to an extent of 25.7%. The water productivity improvement was 24.2% higher than the conventional flooding method. Both the yield reduction and water saving were much higher under Severe AWD (22% and 35%, respectively) than that of mild AWD. However, the paper noted that the reduction in water use, as found by the studies, is because of the reduction in deep percolation (Carrijo, Lundy, & Linquist, 2017). This indicates that the studies considered for the meta-analysis considered the total water applied to the crop and NOT evapotranspiration from the paddy field. Given these trends in yield and applied water use, the economic viability of such methods for paddy growers needs to be carefully examined. Nalley, Linquist, Kovacs, and Anders (2015) investigated the economic viability of different AWD treatments and found the lowest profit in the treatment with the highest water productivity. Considering the fact that both water and electricity are free in countries like India, higher water productivity resulting from reduced water use (with a resultant reduction in yield), means lower net income for the farmers in those areas, as gross income decreases with the reduction in yield and the input cost doesn't change (Kumar & van Dam, 2013).

10.3 The conceptual framework

The conceptual framework presented in Table 10.1 for identifying the nature of interventions for improving water productivity in agriculture is adapted from Kumar (2018) by drawing upon the knowledge available from several published scientific literature on the topic, particularly, the

Table 10.1 Conceptual framework for improving water productivity at plant, plot, system, basin, and regional level: for different basin Water budgets scenarios.

Basin water budget (inflows and Outflow)	Physical, socioeconomic conditions	Basin water demand–supply balance	Basin water supplies vs inflows	Interventions for water productivity improvements				Water productivity trajectory
				Plant level	Plot level	System level	Basin level	
Catchment outflows less than the Inflows ($ET_a + E < P$)	Sub-humid regions—medium to high and very high rainfall. Lack of resources to invest in well irrigation due to poverty	Basin water demand exceeding the supplies (PET + PE > S). Land receiving inadequate irrigation due to high cost of irrigation water. Land lying fallow during summer	Basin water supplies less than the Inflows (abundance of untapped groundwater)	• Waterlogging and salinity resistant varieties	• Supplementary irrigation for rain-fed crops. Intensifying cropping with irrigation expansion. High valued vegetables with expensive irrigation	• Investments in water & electricity infrastructure	Utilization of untapped groundwater resources	Improvement in plot, system, basin and regional level WP. Less groundwater outflows
Catchment outflows less than the Inflows ($ET_a + E < P$)	Semi-arid regions—medium to high rainfall, hilly topography. Terrain conditions makes investment in irrigation prohibitively expensive	Basin water demand exceeding the supplies (PET + PE > S). Land lying fallow during winter and summer	Basin water supplies less than the inflows (Still have unutilized stream flows)		• Supplementary irrigation of rain-fed crops. • Intensifying cropping with irrigation	Investment in water lifting and irrigation infrastructure. Use of return flows from surface irrigation systems through wells and river/drain lifting	Utilization of untapped surface water resources for irrigation expansion in the basin	Improvement in plot, system, basin and regional level WP. Less unutilized stream flows
Outflows equal to Inflows ($ET_a + E = P$)	Semi-arid regions—low to medium rainfall. Less amount of arable land	Basin water supplies equals the demand (S = PET + PE). No land is left unirrigated	Entire inflow is tapped; no additional renewable water in the catchment; Surface water is also imported	Varieties with high genetic yield potential. Varieties with high transpiration efficiency. Varieties with low LAI	Crops with low E/T ratio; crops with improved harvest index. Water-efficient, high valued crops (higher Kg/ ET and Rs/ET)	Reduced diversion of water from the irrigation systems for crop production	Reduced use of water in agriculture. Greater allocation of water from the system for water-efficient crops. Growing water-intensive crops in areas with lower ET demand	Improvement in plot level water productivity. Improvement in WP at system level and basin level. Freeing some water from agriculture

Interventions for water productivity improvements

Basin water budget (inflows and Outflow)	Physical, socioeconomic conditions	Basin water demand–supply balance	Basin water supplies vs inflows	Plant level	Plot level	System level	Basin level	Water productivity trajectory
Outflows equal to Inflows $(ET_a + E = P)$	Semi-arid regions, with low to medium rainfall; Hard rock regions with no groundwater stock	Basin water demand exceeding the supplies $(PET + PE > S)$; Significant chunk of the land lying fallow during nonmonsoon period	Entire Catchment Inflow is tapped; no additional renewable water to harness	Varieties with high genetic yield potential and high transpiration efficiency; Varieties with low LAI; Drought resistant varieties	Crops with low E/T ratio; crops with improved harvest index; Higher Kg/ET through deficit irrigation; Reducing NRDP losses through drip for row crops; Reducing SE through mulching for row crops; High valued crops with higher income (Rs)/ET	Canal lining and other conveyance efficiency improvement measures in deep water table areas	Expansion in irrigated area in the basin using the water saved through efficient irrigation technologies	Improvement in plot, system and basin level water productivity; No basin level water saving
Outflows more than inflows $(ET_a + E > P)$	Semi rid regions with low to medium rainfall; Deep alluvial aquifers with significant groundwater stock; Deep Water Table conditions	Basin water demand met from supplies $(PET + PE = S)$; No arable land left unirrigated (Example: Punjab and Haryana)	Entire catchment Inflow and imported water is utilized; Deficit is met through aquifer mining	Varieties with high genetic yield potential and high transpiration efficiency; Drought resistant varieties	Crops with low E/T ratio; crops with high harvest index; Drips irrigation to reduce NRDP; Plastic mulching for row crops to reduce SE; Introduce new, high value crops with higher income (Rs)/ET	Canal lining; Lining of conveyance channels for irrigation wells	Expansion in irrigated area in the basin using the water saved water	Overall improvement in system level WP; Overall improvement in basin-level WP; Reduced groundwater mining
Inflows far higher than outflows $(P > ET_a + E)$	Humid and sub-humid regions with steep slopes, poor arable land availability	Entire basin water demand met from Supplies $(PET + PE = S)$; West flowing rivers of Western Ghats; Rivers of North East	Catchment Inflow far Exceeding Supplies		Transfer water to water-scarce basins; Introduce water intensive crops which are also water-efficient (Rs/ET)	Infrastructure for transfer of water from water-abundant basins to water-scarce basins	Overall increase in utilizable water resources from the basin	Basin water economy improves; Improvement in regional water productivity

Terms: Inflows; Outflows; genetic yield potential; transpiration efficiency; drought resistant variety; waterlogging and salinity resistant variety; harvest index; evaporation to transpiration ratio. LAI, leaf area index; PET, potential evapotranspiration; PE, potential evaporation; NRDP, nonrecoverable deep percolation; SE, soil evaporation.
Source: adapted from Kumar, M.D. (2018). Water policy, science and politics: An Indian perspective. Elsevier Science, Amsterdam, pp. 326.

definitions of different components of water use by Allen, Willardson, and Frederiksen (1997); the water accounting framework of Molden, Murray-Rust, Sakthivadivel, and Makin (2003) and Molden et al. (2007); the distinction made between consumptive and nonconsumptive fraction, by Perry (2007); and description of physical and environmental conditions for water-saving through microirrigation systems by Kumar, Turral, et al., (2008), Kumar and van Dam (2013), and Kumar (2016).

It also extensively uses knowledge available from a myriad of scientific studies, particularly on the following: manipulating climate to improve crop water productivity (Abdullaev & Molden, 2004; Zwart & Bastiaanssen, 2004); the impact of deficit irrigation on water productivity in kg/ET (Deng, Shan, Zhang, & Turner, 2006; Zhang, 2003); techniques to reduce evaporation/ET ratio, reduce LAI and improve harvest index so as to improve field-level water productivity of crops in kg/ET (Cooper, Keatinge, & Hughes, 1983; Siddiqui et al., 1990; Tanner & Sinclair, 1983); techniques to enhance "transpiration efficiency" to improve plant-level water productivity (Aggarwal, Kalra, Bandyopadhyay, & Selvarajan, 1995; Schmidhalter & Oertli, 1991); and approaches to improve water productivity of crops in economic terms ($/m^3 of water) (Kumar, Singh, et al., 2008; Kumar and van Dam, 2009).

To begin with, the development of such a framework considers the basin as the unit of analysis to explore opportunities for WP improvement. It identifies the interventions that are possible at various scales, i.e., plant-level, plot-level, irrigation system level, and basin level. While doing this, it considers six major situations which are encountered in terms of the basin water budget, as ideally, the interventions will have to be different for different situations. The outcomes of these interventions in terms of the likely changes in WP at various levels are also discussed. All these six situations are encountered in the South Asian region. They are:

- **Case 1**: Subhumid regions with medium to high or very high rainfall/precipitation: basin water demand exceeds the supplies, but supplies less than the inflows (groundwater abundant basins); therefore, the outflows are less than the inflows ("open basins")
- **Case 2**: Semiarid regions, with medium to high rainfall/precipitation and hilly/undulating topography; basin water demand exceeding the supplies; but supplies less than the inflows (surface water abundant basins); therefore, the outflow is less than inflows ("open basins")
- **Case 3**: Semiarid regions with low to medium rainfall/precipitation; basin water supplies equal the demand; entire inflow is tapped and in addition,

surface water is imported into the basin for meeting the demands; outflows equal inflows; entire arable land is irrigated ("closed basins")
- **Case 4**: Semiarid regions with low to medium rainfall/precipitation; hard rock geology with no groundwater stock; basin water demand exceeding the supplies; entire inflow is tapped and no additional water to tap; therefore, the outflow is equal to inflows; a substantial amount of land lying unirrigated ("closed basins")
- **Case 5**: Semiarid regions with low to medium rainfall; deep alluvial aquifers; demand is met through additional supplies from imported surface water and mining of groundwater, along with annual inflows ("closed basins" with severe water stress)
- **Case 6**: Humid and subhumid regions, with very high rainfall; hilly topography; basin water demand is met from supplies; inflows far higher than the supplies (open basins)

Here, the annual inflows include surface runoff and renewable groundwater recharge from precipitation. The outflows include the beneficial and non-beneficial evapotranspiration and nonconsumptive and nonrecoverable "loss." The supplies will have to come from inflows, mining of groundwater stock, water imports, or carryover storage of water from reservoirs. Now the type of water productivity improvement measures possible in a basin would change according to the situations mentioned above. For instance, in a "closed basin," where the basin outflow is equal to inflows, increased utilization of water for irrigation for supplementary irrigation in a domain will not be possible, without affecting the downstream. Similarly, in a humid or subhumid area with shallow aquifers, water productivity improvements through an improvement in conveyance efficiency and water application efficiency will not be possible at the system level, though the productivity of applied water could improve at the plot level.

Whereas in an "open basin" in semiarid regions with medium to high rainfall areas, it would be possible to expand the area under irrigation or provide supplementary irrigation of rain-fed crops with further utilization of the untapped water, thereby increasing WP at field level, and also at the basin level. Here, the nonconsumptive, nonrecoverable water, that goes into the natural sink, is captured and utilized for beneficial transpiration by increasing the transpiration of rain-fed crops or by raising irrigated crops in the land lying fallow during the winter and summer seasons.

In a hot and semiarid or arid area with deep water table conditions ("closed basins with severe water stress"), if the outflows exceed the inflows, improvement in WP at the plant level would be possible through the introduction of varieties with high genetic yield potential and transpiration efficiency. But this can increase the WP at all subsequent levels as well, even without introducing any interventions at the plot, system, and basin levels. Nevertheless, further improvements in WP at the subsequent levels (plot, system, and basin level) would be possible through the selection of crops with a high T/ET ratio, crops with improved harvest index, deficit irrigation, etc. The lining of canals would help save some water in such environmental conditions, at the system level. Further expansion in an irrigated areas in the basin using the saved water would increase the basin level water productivity too, though it won't save any water.

10.4 Findings and conclusions

Over the past two decades, there has been a significant advancement in the science of irrigation and knowledge about the flow of water through unsaturated media. Particularly, scientific knowledge about the soil water balance in irrigated paddy fields and seepage from canals have advanced with the development of scientific models and engineering research in these fields (source: based on Singh, 2005; Kim et al., 2009; Watt, 2008). Yet, irrigated paddy is considered to be a highly water-intensive crop in the South Asian region where it is extensively cultivated, based on the total volume of water that farmers use for irrigating this crop. Efficiency in public gravity irrigation systems is considered very low, based on the proportion of water supplied from the reservoirs that reach the farmers' fields.

As a result, there is continued interest in engineering measures, such as the lining of canals and improved methods of irrigating paddy, as water-saving interventions. Part of the reason is the lack of ability among the professionals involved in such works, whose orientation is mainly towards the construction of civil engineering infrastructure, to assess the complex physical processes governing consumptive water use in irrigation, canal seepage, the reusable fraction from seepage, and real water saving in irrigation. Another reason for the enthusiasm could be the large funds involved in the execution of such schemes. In this chapter, we have discussed two very common foleys in irrigation modernization projects such as canal lining and alternate efficient irrigation in paddy viz., alternate wetting and drying (AWD).

We have also presented an analytical framework for understanding the basin conditions and to identify which among the following interventions for improving water productivity in agriculture such as efficient irrigation, water harvesting, supplementary irrigation, etc., would work for achieving water conservation at various scales. The analytical framework uses the following key variables: (i) renewable water availability in the basin; (ii) the proportion of the renewable water sources already tapped for meeting the current demands; (iii) the proportion of the current demands that are met from the current supply potential; and (iv) the extent of unmet demands in the basin which can be met from further exploitation of the resources.

The analytical framework suggests that if renewable water availability in a basin is higher than the aggregate demand for water in the basin, and yet some of the water needs are compromised due to inadequate supplies, then it is important to invest in water resources development to fill the gap between the demand and the supplies before other options such as improving the productivity of use of the supplied water are explored.

If the demand for water in a basin exceeds the available water supplies, and if the available water resources of the basin are already exploited for meeting various competing demands, then measures such as the control of nonbeneficial evaporation from irrigated fields and reservoirs, transpiration efficiency improvement for plants (crops), etc., can be explored to improve the overall productivity of water use at the basin level, whereby the saved water can be diverted for either expanding the area under crop cultivation or meeting the needs of other sectors of water use.

If the demand for water from various sectors in a basin is met from the existing supply systems, and if there is the mining of water from groundwater stock (or import of water from outside) in addition to the tapping of available renewable water resources, then measures such as the control of nonbeneficial evaporation of water from irrigated fields and reservoirs, reduction in nonrecoverable deep percolation of water from irrigated fields and transpiration efficiency improvement in plants (crop) can be resorted to, to enhance the overall productivity of water use at the basin level, whereby mining of groundwater can be reduced to arrest depletion or the dependence on water import can be reduced.

Instructions on the use of the framework: In order to decide on which one of the above six categories a given basin falls (i.e., whether open or closed), it is important to estimate the water demand, which includes the potential process evapotranspiration (beneficial consumptive water demand) and potential evaporation (PE) (nonbeneficial consumptive water demand) and nonrecoverable losses (deep percolation into unsaturated zones of aquifers, percolation into saline formations and outflows into natural sinks) against the maximum supplies that can be tapped from inflows. For estimating the potential beneficial evapotranspiration in agriculture and forestry (PET), it is necessary to have the hydrometeorological data (solar radiation, temperature, relative humidity and wind speed) and data on potential area that can be brought under different irrigated crops, the seasons in which they are raised and their duration.[c] The other consumptive use sectors include manufacturing process, domestic (rural and urban) and livestock. The potential sources of nonbeneficial evaporation (PE) include swamps, barren soils in irrigated and rain-fed crop land, which remain wet, and reservoirs and open channels for water conveyance. This would again be determined by the hydrometeorological conditions. Similarly, the inflows from precipitation (surface runoff, base flows, and recharge to groundwater) also will have to be estimated. In order to estimate the "outflows" from the basin (depletion), the actual area under irrigated crops and under different types of other vegetation will have to be used.

[c] Modified Penman method can be used to estimate PET for different crops per unit area (in mm) using the hydro-meteorological data.

References

Abdullaev, I., & Molden, D. (2004). Spatial and temporal variability of water productivity in the Syr Darya Basin, central Asia. *Water Resources Research*, *40*(8).

Aggarwal, P. K., Kalra, N., Bandyopadhyay, S. K., & Selvarajan, S. (1995). A systems approach to analyze production options for wheat in India. In J. Bouma, et al. (Eds.), *Systems approach for sustainable agricultural development*. The Netherlands: Kluwer Academic Publishers.

Ahmad, M. U. D., Masih, I., & Turral, H. (2004). Diagnostic analysis of spatial and temporal variations in crop water productivity: A field scale analysis of the rice-wheat cropping system of Punjab. *Journal of Applied Irrigation Science*, *39*(1), 43–63.

Ali, M. H. (2011). Water conveyance loss and designing conveyance system. In *Vol. 2. Practices of irrigation & on-farm water management* (pp. 1–34). New York, NY: Springer.

Allen, R. G., Willardson, L. S., & Frederiksen, H. (1997). Water use definitions and their use for assessing the impacts of water conservation. In J. M. de Jager, L. P. Vermes, & R. Rageb (Eds.), *Sustainable irrigation in areas of water scarcity and drought* (pp. 72–82). England: Oxford. September 11–12.

Carrijo, D. R., Lundy, M. E., & Linquist, B. A. (2017). Rice yields and water use under alternate wetting and drying irrigation: A meta-analysis. *Field Crops Research*, *203*, 173–180.

Central Ground Water Board. (2014). *Ground water year book 2013–14*. Faridabad: Ministry of Water Resources.

Central Water Commission. (2008). Ministry of Water Resources, River Development and Ganga Rejuvenation. *Theme paper-integrated water resources development and management*. 6 p.

Central Water Commission. (2017). Reassessment of water availability in India using space inputs. In *Basin planning and management organisation*. New Delhi: Central Water Commission. October 2017.

Chaba, A. A. (2019). *Reforming agriculture: When power is worth saving.* The Indian Express. 4 January. Available at: https://indianexpress.com/article/india/reforming-agriculture-when-power-is-worth-saving-5520945/. Accessed: 30 July 2019.

Cooper, P. J. M., Keatinge, J. D. H., & Hughes, G. (1983). Crop evapotranspiration—A technique for calculation of its components by field measurements. *Field Crops Research*, 7, 299–312.

Deng, X. P., Shan, L., Zhang, H., & Turner, N. C. (2006). Improving agricultural water use efficiency in arid and semiarid areas of China. *Agricultural Water Management*, 80(1–3), 23–40.

Down to Earth. (2015). *Water use is excessive in rice cultivation.* Down to Earth. July 04, 2015.

Fischer, G., Hizsnyik, E., Prieler, S., & Wiberg, D. (2011). *Scarcity and abundance of land resources: Competing uses and the shrinking land resource base.* SOLAW Background Thematic Report—TR02 FAO: Rome, Italy.

Hira, G. S., & Khera, K. L. (2000). Water resource management in Punjab under rice-wheat production system. *Research Bulletin*, 1, 2000.

Howell, T. A. (2003). Irrigation efficiency. In *Encyclopedia of water science* (pp. 467–472). New York: Marcel Dekker.

Jadhav, P. B., Thokal, R. T., Mane, M. S., Bhange, H. N., & Kale, S. R. (2014). Conveyance efficiency improvement through canal lining and yield increment by adopting drip irrigation in command area. *International Journal of Innovative Research in Science, Engineering and Technology*, 3(4), 120–129.

Kim, H. K., Jang, T. I., Im, S. J., & Park, S. W. (2009). Estimation of irrigation return flow from paddy fields considering the soil moisture. *Agricultural Water Management*, 96(5), 875–882.

Kumar, M. D. (2018). *Water policy, science and politics: An Indian perspective* (p. 326). Amsterdam: Elsevier Science.

Kumar, M. D. (2016). Water saving and yield enhancing micro irrigation technologies in India: Theory and practice. In P. K. Viswanathan, M. D. Kumar, & A. Narayanamoorthy (Eds.), *Springer series in business and economics. Micro irrigation Systems in India: Emergence, status and impacts* (pp. 13–36). Singapore: Springer Singapore.

Kumar, M. D., Bassi, N., & Singh, O. P. (2020). Rethinking on the methodology for assessing global water and food challenges. *International Journal of Water Resources Development*, 36(2–3), 547–564. https://doi.org/10.1080/07900627.2019.1707071.

Kumar, M. D., & Singh, O. P. (2005). Virtual water in global food and water policy making: Is there a need for rethinking? *Water Resources Management*, 19(6), 759–789.

Kumar, M. D., Singh, O. P., Samad, M., Purohit, C., & Didyala, M. S. (2008). Water productivity of irrigated agriculture in India: Potential areas for improvement. In *Proceedings of the 7th annual partners' meet of IWMI-Tata water policy research program, International Water Management Institute, South Asia Regional Office, ICRISAT Campus, Hyderabad, 2-4 April 2008.*

Kumar, M. D., Turral, H., Sharma, B. R., Amarasinghe, U., & Singh, O. P. (2008). Water saving and yield enhancing micro-irrigation technologies in India: When and where can they become best bet technologies? In *Proceedings of the 7th annual partners' meet of IWMI-Tata water policy research program, International Water Management Institute, South Asia Regional Office, ICRISAT Campus, Hyderabad. 2–4 April 2008.*

Kumar, M. D., & van Dam, J. C. (2013). Drivers of change in agricultural water productivity and its improvement at basin scale in developing economies. *Water International*, 38(3), 312–325.

Meijer, K., Boelee, E., Augustijn, D., & van der Molden, I. (2006). Impacts of concrete lining of irrigation canals on availability of water for domestic use in southern Sri Lanka. *Agricultural Water Management*, 83(3), 243–251.

Molden, D., Murray-Rust, H., Sakthivadivel, R., & Makin, I. (2003). A water-productivity framework for understanding and action. In J. W. Kijne, R. Barker, & D. J. Molden (Eds.), *Vol. 1. Water productivity in agriculture: Limits and opportunities for improvement* CABI Publishing.

Molden, D., Oweis, T. Y., Pasquale, S., Kijne, J. W., Hanjra, M. A., Bindraban, P. S., … Hachum, A. (2007). *Pathways for increasing agricultural water productivity.* H040200, International Water Management Institute.

Nalley, L., Linquist, B., Kovacs, K., & Anders, M. (2015). The economic viability of alternative wetting and drying irrigation in Arkansas rice production. *Agronomy Journal, 107*(2), 579–587.

Perry, C. (2007). Efficient irrigation; inefficient communication; flawed recommendations. *Irrigation and Drainage: The Journal of the International Commission on Irrigation and Drainage, 56*(4), 367–378.

Prasad, A. V. (2018). *India: Direct delivery of electricity subsidy for agriculture an innovative pilot in Punjab.* Presented at Geneva, October 30, 2018. http://esmap.org/sites/default/files/KEF%20Geneva%202018/Energy%20Subsidy%20Reform%20Forum%20%202018_Punjab_Final%20-%2025.10.2018.pdf.

Schmidhalter, U., & Oertli, J. J. (1991). Transpiration/biomass ratio for carrots as affected by salinity, nutrient supply and soil aeration. *Plant and Soil, 135*, 125–132.

Sheng, Z., King, J. P., Aristizabal, L. S., & Davis, J. (2003). Assessment of water conservation by lining canals in the Paso Del Norte region: The Franklin Canal case study. In *World water & environmental resources congress 2003* (pp. 1–7).

Siddique, K. H. M., Tennant, D., Perry, M. W., & Belford, R. K. (1990). Water use and water use efficiency of old and modern wheat cultivars in a Mediterranean-type environment. *Australian Journal of Agricultural Research, 41*(3), 431–447.

Singh, R. (2005). *Water productivity analysis from field to regional scale: Integration of crop and soil modeling, remote sensing and geographic information.* Doctoral thesis Wageningen, The Netherlands: Wageningen University.

Tanner, C. B., & Sinclair, T. R. (1983). Efficient water use in crop production: Research or re-search? In *Limitations to efficient water use in crop production* (pp. 1–27). Madison, WI: American Society of Agronomy.

Watt, J. (2008). *The effect of irrigation on surface-ground water interactions: Quantifying time dependent spatial dynamics in irrigation systems.* Doctoral dissertation Charles Sturt University.

World Bank. (2010). Scoping strategic options for development of the Kabul River basin: A multi sectoral decision support system approach. In *Sustainable development department, South Asia region* The World Bank. Report # 52211.

Zhang, H. (2003). Improving water productivity through deficit irrigation: Examples from Syria, North China Plain and Oregon, USA. In J. W. Kijne, R. Barker, & D. J. Molden (Eds.), *Vol. 1. Water productivity in agriculture: Limits and opportunities for improvement.* Cabi.

Zwart, S. J., & Bastiaanssen, W. G. (2004). Review of measured crop water productivity values for irrigated wheat, rice, cotton and maize. *Agricultural Water Management, 69*(2), 115–133.

CHAPTER 11

Water productivity, water use, and agricultural growth: Global experience and lessons for future

11.1 Summary

In this book, we have argued that discussions on water productivity improvements should focus on regions where either water is becoming a limiting factor (either physically or in financial terms) for increasing agricultural production, or due to intensive use in agriculture, less amount of water is available for economic activities and ecological functions. The discussions can leave out regions where land is a limiting factor for crop production. We further argued that the discussion should also leave out regions where, largely, water in the soil profile is used for crop production, and there is a large extent of arable land being put to cultivation and animal grazing, creating an agricultural surplus.

On the global scale, regions that are really "water-scarce" should be distinguished from those that are "food-insecure." The focus of research and investments on water productivity improvements should be on water-scarce regions. This will help save the precious resources being spent on crop and irrigation technologies for improving water productivity. Leaving the issue of finance aside, theoretically, many interventions to improve water productivity including deficit irrigation and adoption of certain shorter duration varieties come at the cost of yield reduction, and can create more problems for the food insecure regions such as eastern India and Bangladesh that are land-scarce. Yield enhancement should be the research priority in the land-scarce regions.

This does not mean that water productivity should not be a concern for food-insecure countries. Within those countries, there are many where available water resources are not yet appropriated fully to intensify cultivation in the available cropland. They need to be separated out and focused on. Examples are some countries in Sub-Saharan Africa such as Ethiopia. They might have to focus on water productivity improvements in crop production, along with investments in irrigation in the long run, as water resources get increasingly appropriated and water availability for ecological functions increasingly become scarce.

In Chapter 2, we have highlighted the need to distinguish food security challenges of countries from that of water security by showing that high renewable water availability does not guarantee agricultural surplus or food self-sufficiency on the one hand, and poor water renewable water availability does not indicate a food shortage. The chapter suggested a methodology for assessing food security and agricultural production potential of countries that took into account arable land availability and climatic conditions along with renewable water resources, by introducing a composite index called "water-land index," which is the multiple of two indices, viz., "water adequacy index" and "per capita cropped area."

Water adequacy index was defined as the ratio of per capita renewable water resources and the multiple of reference evapotranspiration (for 300 days) and per capita cultivated land, and the maximum value for water adequacy index was kept as 1.0, meaning a value higher than this would not result in higher agricultural production potential. A higher value of the multiple of these two coefficients ensured higher agricultural production potential. The composite index of the water-land index was estimated for 130 countries. Another composite index called "water-land-pasture index," which factors in the per capita pasture land availability and the quality of that land into the water-land index was also developed. Its values were computed for all the same 130 countries. Both the indices were validated as the relationship between the water-land index and virtual water export, and water-land-pasture land index and per capita milk production were tested to be strong, statistically.

We found that when the criteria for assessing the magnitude of food security and water management challenges factor in the role of agricultural land, particularly the cultivated land, we have four different categories of countries emerging: (1) countries with a large amount of renewable water and cultivated land, having both water and food self-sufficiency; (2) countries having large amounts of renewable water resources, but also having a disproportionately larger amount of cultivated land resulting in low "water-adequacy," but "food surplus," though facing occasional water shortages; (3) countries having sufficient amounts of cultivated land, but low water availability, and facing different degrees of water shortages and food self-insufficiency; and (4) countries having high values of "water adequacy" most of the time because of large amounts of renewable water and low per capita cultivated land, and sometimes because of a disproportionately lower amount of cultivated land than what the available water can bring under intensive production, combined with low water availability, but mostly dependent on food imports. Availability of rich pasture land along with cultivated land would significantly influence milk production, an important component of food and nutritional security.

For the first set of countries (such as Argentina, Canada, and Brazil), neither water productivity nor biomass output per unit of land is a serious concern, though they may invest in improving both if that helps increase the profitability of farming. For the second set of countries (such as Australia), improving water productivity in economic terms may be of interest as it would help them spread the available water to bring larger areas under irrigated farming, thereby increasing the total farm output in value terms and net income. The third set of countries (such as Pakistan, Israel, Iran, etc.) should be seriously concerned about improving water productivity in both physical and economic terms. While doing this, they should also try and limit the area under crop production so as to cut down on aggregate water use in agriculture in order to free a significant portion of the water for non-agricultural uses in the future. For the fourth category of countries, improving biomass output per unit of land should be the concern.

Chapter 3 discusses the conceptual issues in defining the terms "water use efficiency" and "water productivity." These conceptual issues are because of the definitions not being really holistic, and each one of them having the potential to convey different meanings to a different audiences, (plant physiologists, agronomists, farmers, irrigation engineers, etc.) and in different contexts—from field to river basin—. New definitions of these terms have evolved in an attempt to capture complex situations and explain what parameters are used in the numerator and the denominator for estimating them. While the concept of water productivity is more recent, there is rich international debate about different aspects of water productivity, starting from the usefulness of the concept of water productivity to the determinants of water productivity; and the scale of measurement to the methods of measurement of water productivity to the water-saving impacts of water productivity improvement measures in agriculture. The discussion has become rich with the introduction of the concepts of total factor productivity in agriculture, and the production function, marginal and average water productivity, and allocative efficiency. These concepts have helped identify the new opportunities for as well as limits to improving water productivity in agriculture.

Chapter 4 dealt with the methodologies for measurement/estimation of various water productivity functions at various scales—plant, plot/field, farm, system, and river basin—, and the indicators for measuring the performance of crops and agriculture with regard to changes/improvement in water productivity at these scales. It also covered an extensive review of the scientific studies that dealt with the assessment of water productivity at various scales, including the studies that assessed the impact of technologies (mulching) and climate change on the water productivity of crops.

The chapter reported that agricultural water productivity could be measured at the plant level, plot/field level, farm level, system–level, and basin level, and the scale of assessment depends on the objective and target audience. With the change in the scale of assessment, the determinants used in the denominator and numerator will change. For plant level measurement of water productivity, the determinants that matters is biomass output and transpiration, whereas, for basin level assessment, it should be the economic value of the social, economic, and environmental benefits produced from the use of water against the total amount of water depleted which includes the total amount of water depleted in crop production plus the water that goes into the natural sink. The methodologies for measuring/assessing water productivity vary with scale. For plant level assessment, the methodology involves estimating ET and then partitioning it into transpiration and evaporation. Field level assessment of water productivity involves the measurement of ET using lysimeters or estimation of ET using weather data, and then the measurement of crop yield. The effect of the "unit" chosen for the analysis on the results of water productivity assessment and its implications for water management needs to be understood.

With the change in the determinants of water productivity, the mechanisms for monitoring the improvement in water productivity will have to be chosen carefully, in accordance with the parameters that truly capture the improvement. In the case of improvement in basin-level water productivity, the determinants of water productivity are the economic value of the agricultural outputs and the total amount of water depleted for agricultural production. For basin-level water depletion, the best alternative is to monitor the change in the annual discharge into the natural sink, the amount of water available for other sectors of the economy, and the annual change in the storage of water in the aquifers, etc. can be indicators.

Chapter 5 reviewed the past growth trends in agricultural water productivity globally and regionally, taking a 40-year time period. However, the analysis is confined to certain broad indicators of the resultant changes at the macrolevel such as rate of change in yield of major crops, rate of change in the value of agricultural outputs, and change in total water withdrawal for agriculture.

Worldwide, annual growth in crop production has been mainly attributed to yield improvements, followed by arable land expansion, and an increase in cropping intensity, but with a remarkable variation in the extent of contribution of different factors across regions. In the developed countries, the modest growth in agricultural output was mainly from a substantial increase in crop yields, with the area under crop cultivation and cropping intensity actually declining. In developing countries, yield gain has contributed majorly to crop output growth, the contribution of intensified cropping and arable land expansion is also significant. Contributions of arable land expansion and increase in cropping intensity are very high in Sub-Saharan Africa, whereas in the case of South Asia, the contribution of increased cropping intensity is significant. Irrigation water withdrawals are also excessively high in South Asia, with the highest in near east/north Africa. Irrigation water withdrawal is still very low in Sub-Saharan Africa.

The overall economic value of agricultural outputs has also increased globally, at a CAGR of 2.2% during 1961–2007. The growth rate was only marginally higher in South Asia (2.9%). In Sub-Saharan Africa, the growth rate was lower (2.6%). The growth rate was highest in East Asia (4%). The historical trends in the value of key determinants/drivers suggest that water productivity in crop production has increased globally and more so in developing countries. The highest improvement in water productivity is expected to have occurred in East Asia, which had recorded a remarkable increase in crop yields without a substantial increase in irrigation water withdrawal. The increase in crop outputs, the extent of contribution of yield growth to this growth, and the increase in the value of agricultural outputs suggest an increase in crop water productivity in South Asia over time. However, the rate of improvement (in physical and economic productivity of water in crop production) is unlikely to be as high as that of crop outputs or value of these outputs, respectively, as a quantum jump in irrigation water withdrawals in these countries might also have contributed to yield growth.

In Chapter 6, we reviewed the situation in different regions of the world with regard to the future possibilities for raising crop yields, expanding arable land, increasing irrigation potential, or intensifying cropping, and improving water productivity for meeting agricultural growth challenges. As regards the future demand for agricultural outputs, it is expected to grow due to population growth, and increasing per capita calorific requirements as a result of the increasing purchasing power of the people in most regions. The growth in demand is estimated to be 1.10% at the global level. However, major variations in growth trends between regions are expected due to the difference in population growth trends and the growth rate in per capita calorie intake. The demand growth is likely to be lowest in the developed countries owing to the lowest or negative population growth rates and the lowest annual growth rate in average calorie intake. The projected annual growth rate in calorie intake is highest for Sub-Saharan Africa, which currently has the lowest average calorie intake, followed by South Asia and East Asia. As a result, the growth rates in aggregate demand for agricultural outputs are high for these regions. Consequently, the projected growth in demand for agricultural commodities is 2.40% for S. S. Africa, 1.50% for North East/North Africa, and 1.70% for South Asia.

The agricultural outputs are also likely to increase globally at the rate of 1.10% per annum to meet the growing demand, with significant regional variations. The highest output growth is expected in Sub-Saharan Africa (2.40%), followed by South Asia (1.6%). The factors that are expected to contribute to this growth will be different in different regions. In Sub-Saharan Africa, the expansion of arable land along with the creation of irrigation facilities will be a major contributor. In South Asia, the main contributor will be irrigation (0.40% per annum), and a lot of it is projected to be from inter-basin water transfer projects. Increased agricultural production achieved mainly through the expansion of arable land results in an effective increase in the utilization of green water from the natural catchments. This will raise water productivity in agriculture in Sub-Saharan Africa. The increase in irrigation water use can also result in water productivity improvements at both field-scale and basin-scale in S.S Africa and South Asia. This is because when rainfed crops get irrigated, it leads to a disproportionate increase in yield of the crop and net income from crops when compared to the increase in crop consumptive use of water. However, pressure on water resources will increase as a result of these changes, particularly in South Asia, North East and North Africa, and East Asia.

In Chapter 7, we undertook a deeper analysis of agricultural productivity and water productivity trends in the South Asian region. Agricultural productivity had increased dramatically in the region over the past four to five decades, with green revolution marking its beginning. Among the three factors that had contributed to the growth, as discussed earlier, arable land expansion had the smallest effect, due to the limited possibility of growth of cultivated land in the region. Intensification of cropping had the second-largest effect, which again was possible with the creation of irrigation facilities, especially wells. The effect of yield improvement on agricultural growth was the largest. However, it is quite likely that in many areas, the yield improvement is either because of the realization of genetic yield potential with the crop water requirement being fully met by better and efficient use of inputs or by virtue of more farmers adopting high yielding varieties with better inputs due to the availability of irrigation water.

The yield growth achieved for cereals in South Asia over the years is phenomenal (Mughal & Fontan Sers, 2020), with a 214% increase over a 56-year period from 1961 to 2017 (Source: FAOSTAT, 2019). But, its effect on water productivity for those crops is not fully known. However, with the increase in irrigation inputs, the yield and net income from crops had increased substantially, disproportionately higher than the extent to which crop consumptive use in the form of ET had increased. This means, with an increasing volume of water from a basin being appropriated for irrigated crop production, the overall water productivity in agriculture in the basin improves considerably. However, this trend cannot continue once the basins are "closed" when every drop of water generated in the basin annually is used and reused. Then the opportunity for further improvement in water productivity will have to come from a reduction in nonbeneficial E and nonrecoverable deep percolation.

The increasing adoption of drip irrigation and mulching for row crops witnessed in the naturally water-scarce regions of India helps reduce nonbeneficial consumptive and nonconsumptive uses of water. The practice of mulching with drips can convert the nonbeneficial consumptive use into beneficial transpiration, and also reduce nonrecoverable deep percolation. With the area under drip irrigation alone exceeding 3.14 m per ha in the Indian states that are known for water scarcity, some notable improvement in agricultural water productivity must have been witnessed. Further, in the recent past, the region has also witnessed considerably high growth in agricultural output in value terms due to the shift to high-value produce such as fruits, vegetables, and milk. Farmers' preference for high-value fruits and vegetables gives a further push to the adoption of precision irrigation technologies such as drips and mulching.

Contrary to the situation in South Asia, irrigation development remains quite low in Sub-Saharan Africa. Vast areas of cropland even in the semiarid zone are under rainfed production resulting in poor yields, while a large quantum of water in the region's river basins remains untapped owing to inadequate public investments for building water infrastructure. Historical growth in yield of crops has been quite poor, unlike South Asia. With the dominance of subsistence crops, especially cereals, crop diversification is also very low.

Chapter 8 dealt with the measures for improving water productivity in rainfed and irrigated agriculture in South Asia and Sub-Saharan Africa. The chapter began with a description of the several agroecologies that exist in the two regions. The relative availability of water in relation to land also changes from region to region. Irrigated agriculture is practiced in both low rainfall semiarid (and arid), water-scarce regions, and high-rainfall, subhumid water-rich regions. The need for improving water productivity is obviously more in the former.

In this chapter, we have discussed several options for raising water productivity in irrigated agriculture in the two regions. In the case of South Asia, the options fell under the following broad categories: (1) technological change for improving the technical efficiency improvements, (2) other measures that include crop management, crop shifts, taking advantage of the climate to decide on the cropping pattern for different regions, water delivery control, improving the quality and reliability of irrigation; (3) efficient pricing of irrigation water and electricity for affecting improvements in technical efficiency and allocative efficiency; and (4) institutional changes, comprising restrictions on water allocation and water rights. We argued that in several Indian states, Pakistan, and Sri Lanka, the opportunities for affecting technology changes for improving the physical efficiency of water use in agriculture through institutional and policy measures are considerably high.

As regards rainfed agriculture, with the provision of supplementary irrigation, a shift from short-duration crops to long-duration crops would be possible in around 228 Indian districts, increasing both yield and water productivity considerably. The shift from rainfed paddy to irrigated paddy is possible in large areas with the provision of irrigation water. With large-scale water resources development happening in Central India and South-Central India, significant areas under rainfed crops could be brought under irrigated production. In addition, there are long-duration rain-fed fine cereals which have moderate consumptive use of water (300–425 mm) in 117 districts of India. These crops, concentrated in Eastern India and Central India, show very high yield gaps. The use of better crop technologies and better inputs could also result in a significant improvement in water productivity through yield enhancement, which would be the effect of nutrients and proportion of the ET being used up for transpiration (Amarasinghe & Sharma, 2009).

In the context of Sub-Saharan Africa, it was argued that the interventions to promote water productivity improvements should focus on the following: (i) increasing the utilization of available surface water resources for irrigation development in the semiarid (Sahelian) zone and the Savannah; (ii) promoting well irrigation in the command areas of large and medium gravity irrigation systems through economic incentives; (iii) small water harvesting systems in the zones where rainfall is reliable and aridity is less, as a source for supplementary irrigation to high-value crops grown in the closest fields, combined with microirrigation systems; and (iv) promoting microirrigation in well commands to irrigate high-value crops.

Further, an appropriate subsidy structure for promoting microirrigation systems shall be designed so that the capital cost of the system is less than the value of incremental benefits accrued from crop production owing to the system use. The extent of subsidy shall be worked out on the basis of the economic value of the externality induced by the technology on the society through water saving. Hence it is clear that in water-scarce regions, the capital subsidy for efficient irrigation systems such as drips should be kept high in view of the high social benefits accrued through water-saving, and the private cost of the system becomes less than the incremental economic benefit derived from the use of the technology.

After having discussed the various technical and institutional alternatives for improving water productivity (Chapter 8), in Chapter 9, we identified and discussed the constraints to improving water productivity in irrigated and rainfed areas of South Asia and Sub-Saharan Africa. We argued that in the case of rainfed agriculture, the constraints are social, socioeconomic, and financial in nature. The poor people in the backward areas of both the regions with limited access to land and water resources, lack knowledge about technologies and markets and entrepreneurial skills and hence are not able to invest in high productivity farming systems. The regions of South Asia and SS Africa, which are relatively water-rich, have very high concentrations of native tribal people who are used to growing subsistence crops that have low economic returns and water productivity. Hence, most of the pre-condition for achieving water productivity gain through supplementary irrigation is unlikely to be met. Investments for water harvesting and groundwater recharge schemes that can help improve water productivity in rain-fed farming systems are very high when compared to the amount of water they can provide.

Hence, large-scale government financing would be required in the face of the poor ability of the tribes to mobilize financial resources.

As regards irrigated agriculture, the constraints are many more, though most relevant to South Asia. They are physical, technology-related, socio-economic, policy-related, institutional, market-related, and environmental. As regards physical constraints, in many regions of South Asia, where crops can be grown with less amount of water by virtue of the climate, arable land availability is very poor. The yield levels are also lower for major crops as compared to regions that are arider. The lack of ability to introduce technologies that can restrict the access of millions of farmers to water and electricity is a technology constraint for improving water productivity. The policy constraint stems from inefficient pricing of water and electricity supply for irrigation in most South Asian countries. In the absence of an effective institutional regime for restricting farmers' access to groundwater in most situations, the farmers' interest is in maximizing the returns per unit of land, and not maximizing returns per unit of water. The irrigation corresponding to high return per unit of land is usually much less than that which corresponds to high returns per unit of water. Hence, there is a direct conflict between private interests and societal interests. Productivity gains through a shift to high-value crops pose significant market risks. Environmental concerns of improving groundwater sustainability in certain cases, and preventing waterlogging in certain other cases also limit our options to improve water productivity in irrigated agriculture.

As discussed in Chapter 2, there are many regions in the world that are water-abundant, but land-scarce, and scarce in both land and water. Many countries in South and South East Asia belong to the former and some countries in SS Africa belong to the latter. Investing in water productivity in such regions will not make much sense. Hence in order to understand the opportunities and constraints for improving water productivity and to know where such measures would yield benefits in terms of increase in agricultural outputs, rural growth, and poverty reduction, a conceptual framework was developed to characterize the river basins and to suggest measures for water productivity improvements at various scales. This was discussed in Chapter 10.

The framework considered four important attributes to characterize the river basins, as follows: (i) the renewable water resources available in a basin; (ii) how much of the basin's available water resources are appropriated for meeting the existing demands, including agriculture; (iii) what proportion of the current demands are met from the current supply potential; and, finally (iv) whether there are unmet demands in the basin which can be met from further exploitation of the resources. For the application of the framework, it considered the river basin as the unit of analysis. Options are identified at plant scale, field-scale, irrigation system scale, and basin scale. The framework considered six major situations which are encountered in terms of the basin water budget. The outcomes of these interventions in terms of likely changes in WP at various levels are also discussed. All these six situations are encountered in the South Asian region.

In the second section of the chapter, we have discussed some of the most touted, yet flawed ideas implemented in the past in different countries to improve water productivity or to save water in agriculture. Yet, they are being replicated in development projects due to their continuing popularity in policy circles. They are the lining of canals in the command areas and the use of efficient methods of irrigation for paddy. These schemes are still very popular in South Asian countries, particularly, India, Pakistan, Afghanistan, and Sri Lanka, which is evident from the fact that they are used as a straight-jacket solution in every irrigation modernization project.

11.2 Lessons for future

To conclude, from the reviews and analyses presented in the earlier chapters of this book, the following conclusions can be made:

(i) The traditional approach of considering the renewable water resources of a region to assess the extent of the sufficiency of water for agricultural production in that region, without looking at the amount of arable land that is cultivated, will be highly misleading as an indicator to decide on water productivity improvement measures in that sector for managing water demand. A new approach is required to assess the food and water challenges which integrates arable land as a key variable for estimating both the supply and demand of water for food and agricultural production, and production of food and other agricultural commodities (Kumar, Bassi, & Singh, 2020).

(ii) The use of this approach will force us to reestimate the future water demand for agriculture in such regions to include the quantum of surplus food for export to the neighboring food-insecure regions or countries, even if the former is water-scarce. This is quite contrary to the conventional approach of suggesting food import by such regions, which does not take into account the fact that water-rich regions that are supposed to produce this food often lack sufficient arable land to produce food even to meet their own needs. The new approach will also force us to revisit the estimates of future water requirements for various uses for regions that are land-scarce (Kumar et al., 2020). Once we do this, we will know that the former category of regions would require aggressive measures for improving water productivity in agriculture, whereas, the latter may not require any even if the renewable water availability might not be high.

(iii) There is a clear trade-off between maximizing water use efficiency (kg/ET) and maximizing the yield of crops (Fan, Wang, & Nan, 2018). There can also be a trade of between maximizing water productivity (k/m^3) and maximizing yield (Molden, Murray-Rust, Sakthivadivel, & Makin, 2003). Hence, water productivity improvements in agriculture may not always lead to greater benefits for the agriculturists and sometimes can be at the expense of net revenue or income from farming. It will be definitely so in situations where farmers are not confronted with the positive marginal cost of using irrigation water due to which the net, as well as gross returns from farming, increases with yield. For introducing water productivity improvement measures, as a general principle, it is important to target regions experiencing a physical scarcity of water in relation to arable land or regions where the cost of production and supply of water is very high.

(iv) Globally, water productivity in agriculture is constantly increasing as a result of several advancements at the technology front (plant breeding, irrigation management technologies and practices), irrigation expansion (including supplementary irrigation), arable land expansion, and reallocation of water from subsistence crops to high-value cash crops. However, the causes for these changes are different in different regions, depending on the regional differences in the demand for agricultural production growth, advancements in crop sciences, water scarcity, and the opportunities available for arable land expansion.

(v) There is a clear distinction between water productivity improvements and water conservation. Many interventions that result in water productivity improvements even at the basin level, can actually lead to a greater depletion of the available water resources, though might result in disproportionately higher agricultural outputs. One good example is the use of fallow land for expanding cultivation that helps increase the utilization of water in the soil profile which otherwise is lost in nonbeneficial evaporation, for crop production. While it increases the agricultural outputs and overall water productivity at the basin level, it can reduce the runoff in the basin. Another example is the use of untapped water resources in river basins to expand irrigation or to replace rainfed farming with irrigated crops. While this will also increase agricultural outputs substantially, it will leave less amount of water for the environment. Both situations are likely to emerge in Sub-Saharan Africa. A third example is the aggregate impact of the use of efficient irrigation technologies. As noted by many scholars, with the adoption of efficient irrigation technologies, agricultural water productivity improves in the basin (Expósito & Berbel, 2017). But the aggregate water use in a basin may decrease or increase depending on a variety of conditions such as agricultural system characteristics, aquifer characteristics, previous conditions of irrigation infrastructure, and induced changes in crop rotations (Berbel, Gutiérrez-Martín, Rodríguez-Díaz, Camacho, & Montesinos, 2015; Sanchis-Ibor, Macian-Sorribes, García-Mollá, & Pulido-Velazquez, 2015) and conditions with respect to water and power supply (Kumar, Turral, Sharma, Amarasinghe, & Singh, 2008).

(vi) To arrive at the need for and to decide on effective interventions for improving agricultural water productivity at various scales, it is important to use a proper analytical framework. Such a framework involves understanding the characteristics of the basin vis-à-vis the following: (a) the renewable water resources available in a basin; (b) how much of the basin's renewable water resources are appropriated for meeting the existing demands; (c) to what extent the current demands are met from the existing supply potential; and (d) whether there are unmet demands which can be met from further exploitation of the resources. If we consider these attributes, six different situations are possible and all of them exist in South Asia.

(vii) Different water productivity improvement measures—such as the use of plant breeding for improving transpiration ratio (kg/T), manipulating plant architecture to reduce evaporation ratio (E/ET), to use of mulching in the fields to supress soil evaporation, to canal lining in arid regions with deep water table conditions for reducing nonbeneficial, nonconsumptive use and use of evaporation retardants in reservoirs to reduce nonbeneficial consumptive use—, can have different impacts as one moves from the plant to the river basin depending on the climate, physical environment, the extent of land use and degree of water resource development, etc. However, plant breeding to improve transpiration ratio and harvest index are two interventions that can help improve agricultural water productivity at all scales.

(viii) To affect improvements in crop water productivity through technology measures and practices such as deficit irrigation, use of drip irrigation plus mulching, it is necessary to align policies and institutions in the water sector as the desired use of these technologies involve behavioral changes with regard to the use of water. It is more important for South Asian countries such as India, Sri Lanka, Pakistan, and Afghanistan. The institutional and policy changes should be such that they induce opportunity costs of using water. This can be achieved through volumetric pricing of water and the establishment of water entitlements/water rights in the case of surface irrigation, and pro-rata pricing of electricity, energy rationing in the case of groundwater irrigation. In the absence of such regime changes, the farmers would continue to engage in wasteful practices even after adopting efficient irrigation technologies.

References

Amarasinghe, U. A., & Sharma, B. R. (2009). Water productivity of food grains in India: Exploring potential improvements. In M. D. Kumar, & U. Amarasinghe (Eds.), *Strategic analyses of the National River Linking Project (NRLP) of India series 5. Water productivity improvements in Indian agriculture: Potentials, constraints and prospects* (p. 163).

Berbel, J., Gutiérrez-Martín, C., Rodríguez-Díaz, J. A., Camacho, E., & Montesinos, P. (2015). Literature review on rebound effect of water saving measures and analysis of a Spanish case study. *Water Resources Management, 29,* 663–678.

Expósito, A., & Berbel, J. (2017). Agricultural irrigation water use in a closed basin and the impacts on water productivity: The case of the Guadalquivir river basin (Southern Spain). *Water, 9,* 136. https://doi.org/10.3390/w9020136.

FAOSTAT. (2019). *Food and Agriculture Data from FAO for South Asia for 2019.* Food and Agriculture Organization of the United Nations. http://www.fao.org/faostat/en.

Fan, Y., Wang, C., & Nan, Z. (2018). Determining water use efficiency of wheat and cotton: A meta-regression analysis. *Agricultural Water Management, 199,* 48–60. https://doi.org/10.1016/j.agwat.2017.12.006.

Kumar, M. D., Bassi, N., & Singh, O. P. (2020). Rethinking on the methodology for assessing global water and food challenges. *International Journal of Water Resources Development, 36*(2–3), 547–564. https://doi.org/10.1080/07900627.2019.1707071.

Kumar, M. D., Turral, H., Sharma, B. R., Amarasinghe, U., & Singh, O. P. (2008). Water saving and yield enhancing micro irrigation technologies in India: When do they become best bet technologies? In M. D. Kumar (Ed.), *Managing water in the face of growing scarcity, inequity and declining returns: Exploring fresh approaches, Volume 1, proceedings of the 7th annual partners' meet of IWMI-Tata water policy research program* (pp. 13–36). Hyderabad: ICRISAT.

Molden, D., Murray-Rust, H., Sakthivadivel, R., & Makin, I. (2003). A water-productivity framework for understanding and action. Water productivity in agriculture: Limits and opportunities for improvement. In *Comprehensive assessment of water management in agriculture*. UK: CABI Publishing in Association with International Water Management Institute.

Mughal, M., & Fontan Sers, C. (2020). Cereal production, undernourishment, and food insecurity in South Asia. *Review of Development Economics*, *24*(2), 524–545.

Sanchis-Ibor, C., Macian-Sorribes, H., García-Mollá, M., & Pulido-Velazquez, M. (2015). Effects of drip irrigation on water consumption at basin scale (MIJARES RIVER, SPAIN). In *26th Euro-Mediterranean regional conference and workshops, 'Innovate to improve Irrigation performances', 12–15 October 2015, Montpellier, France.*

Index

Note: Page numbers followed by "*f*" indicate figures and "*t*" indicate tables.